T0210838

KEN BURNS'S

America

Gary R. Edgerton

palgrave

for St. Martin's Press

KEN BURNS'S AMERICA
Copyright © Gary. R. Edgerton, 2001.
Softcover reprint of the hardcover 1st edition 2001 978-0-312-23646-5
All rights reserved. No part of this book may be used or reproduced in any manner
whatsoever without written permission except in the case of brief quotations
embodied in critical articles or reviews.

First published 2001 by PALGRAVE™
175 Fifth Avenue, New York, N.Y. 10010.
Companies and representatives throughout the world.

PALGRAVE is the new global publishing imprint of St. Martin's Press LLC
Scholarly and Reference Division and Palgrave Publishers Ltd. (formerly Macmillan
Press Ltd.).

ISBN 978-1-349-63110-0   ISBN 978-1-137-05482-1 (eBook)
DOI 10.1007/978-1-137-05482-1

Library of Congress Cataloging-in-Publication Data

Edgerton, Gary R. (Gary Richard), 1952-
   Ken Burns's America / by Gary R. Edgerton.
      p. cm.
   Includes bibliographical references and index.

   1. Burns, Ken, 1953---Criticism and interpretation. I. Title.

PN1992.5.B79 E34 2001
791.45'0232'092--dc21   2001019451

Design by planettheo.com

First edition: November 2001
10 9 8 7 6 5 4 3 2 1

Contents

Ten pages of photos appear between pages 108 and 109.

Preface and Acknowledgments

Scholars have generally paid less attention to documentary films and television programs than to their fictional counterparts, reflecting both the longstanding priorities of the motion picture and television industries as well as the customary preferences of mass audiences throughout much of the last century. Since the early 1980s, in particular, the conventional boundaries between fact and fiction, information and entertainment, and among film, photography, TV, video, and multimedia production have grown increasingly indistinct with each passing year. Ken Burns became the first breakthrough documentarian of his generation, working mainly in television and reaching tens of millions of people with his prime-time vision of America. Above all else, Burns's influence as a nonfiction filmmaker and a popular historian springs from his close association with public TV, not in spite of this fact. The question of whether Burns will someday make a fiction film is mostly beside the point by now. Contemporary motion picture and television producer-directors regularly consider the full repertoire of nonfictional and fictional stylistics when fashioning their stories for the screen. Burns's success is directly attributable to his willingness to bridge traditional categories in his highly personal (and often imitated) approach to producing history on film for TV.

After a quarter century of work including 16 major television specials (with several more already in preparation), at mid-career Ken Burns is presently public television's most prominent and acclaimed producer-director. *Ken Burns's America*, in turn, is the first book-length study to comprehensively examine this innovative filmmaker as a television auteur, a pivotal programming influence within the industry, and a popular historian who portrays a uniquely singular and compelling version of the country's past. This volume's three-fold agenda, is to, first, delineate his personal influences, his distinctive and well-recognizable style, and the development and maturation of his ideological outlook, identifying those features that make him one of the most significant cultural voices on TV today. Burns is next viewed as the owner and executive producer of his own independent production company operating on the periphery of public television's institutional framework. And, last, he is analyzed as a popular historian who reevaluates the nation's historical legacy from a new generational perspective. This first book-sized assessment of Burns's historical documentaries is not intended in any way as a biography, although biographical elements do inevitably surface as a way of shedding further light on the films under study.

The Civil War is both the watershed event in Burns's career and the impetus for the recent renaissance in documentary programming on American television. This miniseries is therefore the ideal place to start to define and illustrate Burns as a TV producer-director and popular historian. Chapter 1, "Rebirth of a Nation: Reframing *The Civil War* (1990) on Prime-Time Television," introduces Burns after years of refining his skills as an artist working in photography, film, and video as a storyteller and learning the business of television production as owner of an independent company. All of Burns's stylistic elements come together as never before in *The Civil War,* as does his producing talent for shepherding a project of this size, scope, and ambition from initial concept to public release. After September 1990, the month of its release, Burns's public persona is also established. He becomes a national celebrity, which helps his subsequent television career but noticeably strains his relations with the historical community. *The Civil War* asserts in one fell swoop that history is no longer the principal domain of specialists, as it had been for more than half a century, but is now relevant and compelling for everyone—and this time on TV.

Chapter 2, "Life Lessons: Learning the Basics on *Brooklyn Bridge* (1982)," essentially "flashbacks" to the beginning of Burns's professional career in 1975 when he and two of his college friends—Buddy Squires, who is still his principal cinematographer, and Roger Sherman—started their own independent production company, Florentine Films. They struggled for a number of years doing freelance assignments and finishing a few short documentaries before beginning work in 1977 on their first major project, based on historian David McCullough's book, *The Great Bridge* (1972). Four years later, they completed *Brooklyn Bridge,* sponsored partly by the Television Laboratory at the public TV station, Thirteen/WNET, in New York City. This historical documentary won several important commendations, including an Academy Award nomination, thus ushering Ken Burns and Florentine Films into the system that is the Public Broadcasting Service where he and his company have happily remained ever since. Overall, this second chapter closely explores Burns's seminal influences (i.e., the films of John Ford, photographer Jerome Liebling, and David McCullough), his switch from fictional to nonfictional subjects and stylistics, and his evident growth as an artist as illustrated by assessing his first two efforts as a producer-director, *Working in Rural New England* (1976) for Old Sturbridge Village, a living history museum in central Massachusetts, and *Brooklyn Bridge,* which was completed six years later.

Burns becomes more confident and comfortable with the professional choices he's made during the 1982-1988 time frame covered in Chapter 3, "Variations on a Theme: American Originals, Symbols, and Institutions." At the beginning of this period, he has just moved to a quiet, picturesque village on the Connecticut River, Walpole, New Hampshire, which allows him to live inexpensively and work only on those projects that he creates and develops for himself along with his closest colleagues at Florentine Films. This proves to be a prolific stretch of years for Burns and his creative team, resulting in *The Shakers: Hands to Work, Hearts to God, The Statue of Liberty, Huey Long, Thomas Hart Benton,* and *The Congress.* Although at first it appears that he is embracing a wide assortment of subjects, this third chapter delineates the underlying characteristics that hold his approach to the American past together. Burns's work from this point onward demonstrates certain narrative and ideological imperatives that support each other, and which form an image of the nation that is romantic (i.e., based on emotions), liberal pluralist, and largely celebratory in nature. By the mid-1980s, Ken Burns begins to create a series of television specials as morality tales, drawing upon epic events, landmarks, and institutions of historical significance, populated by heroes and villains who allegorically personify certain virtues and vices in the national character as understood through the popular mythology of America's collective memory. This is a stylistic formula that comes to full fruition in *The Civil War,* an approach that he and many other TV documentarists of his generation will continue to develop and refine throughout the rest of the 1990s and into the current decade.

Empire of the Air: The Men Who Made Radio is an important transitional film in Burns's career. Chapter 4, "The Creative Team as Historian: Inside the Production Process on *Empire of the Air: The Men Who Made Radio* (1992)," offers a glimpse into the directions Burns will take in the future. *Empire of the Air* was conceived and created while *The Civil War* was being edited and later released to wide attention and acclaim, a period during which Burns's professional profile changed dramatically as he became a national celebrity virtually overnight. The heightened work environment and reaction surrounding *The Civil War* also affected *Empire of the Air*'s production process in minor ways at first, but then more tangibly after the summer of 1990. The main purpose of Chapter 4 is to analyze *Empire of the Air: The Men Who Made Radio* as made-for-television history. It offers a revealing object lesson on the shared authorship that typically occurs when creating mediated history on film for TV. *Empire of the Air,* moreover, appeared first as a book, then as a major public television special, and finally as a radio play, providing a representative example of how history (like any other narrative genre) commonly is adapted across multiple media forms these days and, in turn, is directed toward varying segments of the American mass audience.

By the release of *Empire of the Air,* "Ken Burns" is becoming a brand name and the market is fundamentally changing for made-for-television histories. Most everything about Ken Burns's professional life is in transition after the unparalleled reception and success of *The Civil War.* Chapter 5, "A Whole New Ball Game: *Baseball* (1994) and *The West* (1996) as Event TV," focuses specifically on Burns as an executive producer and business entrepreneur. *Baseball* is analyzed from its inception through its September 1994 premiere. As *Baseball* mushroomed from nine to eighteen and a half hours, so did Florentine Films/American Documentaries, Inc. in kind. This chapter basically maps out how Burns comes to realize his company's niche in the television business by designing *Baseball* from the outset as "event TV." It first follows the rigors of conceptualizing and producing this miniseries. *Baseball* also required licensing agreements for more than two dozen ancillary products, while at the same time Burns was acquiring future projects and assigning on-the-line producers to develop these new properties according to his own well-established style and approach. The second half of this chapter follows *The West* as a case study in this regard. Ken Burns, most significantly, assumed an ever greater role in this period as executive producer and company figurehead to go along with his usual directorial duties.

Chapter 6, "American Lives: *Thomas Jefferson* (1997) and the Television Biography as Popular History," primarily examines Burns as a historical biographer in his work on the *American Lives* series, co-sponsored by General Motors as part of its "Mark of Excellence" television presentations. *American Lives* debuted in February 1997 with a reported 17 million viewers for *Thomas Jefferson,* and was followed in successive Novembers by *Lewis & Clark: The Journey of the Corps of Discovery* (1997), *Frank Lloyd Wright* (1998), and *Not For Ourselves Alone: The Story of Elizabeth Cady Stanton and Susan B. Anthony* (1999). *Mark Twain* completes the original quintet of *American Lives* episodes, setting the stage for five more installments to begin production in 2002 and to continue in succession over the course of the current decade. Burns's role as a popular historian is specifically highlighted in Chapter 6, using *American Lives* for illustrative examples. The biographical formula is essentially underscored for its potential to simulate powerful feelings of intimacy in audience members as they watch and relate to the featured characters' life stories in the privacy and comfort of their own homes. An interlocking ritual of producing, telecasting, and watching biographies, such as the ones in *American Lives,* is presented as a shared ceremonial experience for the producers and the vast numbers of viewers who tune in to see these newly adapted screen versions of pivotal figures from American history.

Chapters 4, 5, and 6 also examine and underscore Ken Burns as a television auteur, an executive producer, and a popular historian. Chapter 7, "Ken Burns's America Reconsidered: Mainstreaming *Jazz* (2001) for a National Audience," reemphasizes and summarizes the three-dimensional nature of Burns as a TV professional by using a thorough analysis of *Jazz* as its focal point. This final chapter ends with five summary observations about Burns's televisual style and his developing conception of America. Burns's work is very much of the moment. His historical documentaries are generally liberal on social issues, as is evident by his abiding interest in race relations and the country's democratic ideals, while concurrently traditional in respect to core American values. In the summer of 1999, Burns also signed an unprecedented ten-year agreement with General Motors whereby the automaker will underwrite 35 percent of the production costs and 100 percent of the extensive promotional and educational outreach efforts for all of his public television specials throughout the remainder of the decade. Suffice to say, Burns's professional plate is as full and active as ever, while his vision of America—which is very much a work in progress—is destined to remain a fixture on TV screens in the United States and internationally for many years to come.

One of the genuine pleasures of doing a research and writing project of this magnitude is the opportunity it has afforded me to meet and interact with scores of new contacts and acquaintances, all of whom shared information both on and off the record that helped me in incalculable ways. To some degree, it is hard not to name all of the hundreds of people I have spoken with about television and film as history over the past five years—including film and television professionals, scholars and critics, and various librarians and archivists—because all have indeed influenced me in one way or another with their various perspectives and opinions from their own distinctive backgrounds and areas of expertise. Let me begin by thanking those editors who encouraged me to start writing about Ken Burns's films in the first place, including Greg Bush at *Film & History,* Mike Marsden and Jack Nachbar at the *Journal of Popular Film and Television,* and Dick Pack at *Television Quarterly.* Their initial support certainly started me thinking about expanding these preliminary explorations into a far more extended study on the historical documentaries of this highly influential and creative producer-director.

Many other people also helped me during the researching of this book. Of special note, I want to acknowledge the important assistance provided by Michael Taft, John White, and David Camp at the Folklore Archives of the Wilson Library at the University of North Carolina at Chapel Hill where the *Ken Burns Collection* is housed; Jim Dougherty, Virginia Field, Nancy Rogers, Jim Vore, and Laura Nelson at the National Endowment for the Humanities; Camilla Rockwell and Pam Baucom of Florentine Films (at the Walpole, New Hampshire incarnation); Meg Haley of the Visual Resources Library at Old Sturbridge Village; Elizabeth Hitchcock at the University of Pennsylvania; Angela Stangarone and Sarah Botstein of Owen Comora Associates; Joseph DePlasco of Dan Klores Associates; Amy Fletcher, Rosalyn Teichroew, Linda Delgado, and John Morison of WHRO, public television in Hampton Roads; and Karen Yankowski of PBS Video.

Thanks, too, to those who have sent me photographs, audiotapes, videotapes, and various articles and printed materials over the years, including Carolyn Anderson, Larry Bonko, David Culbert, Kathy Fuller, Kathy Merlock Jackson, Michael Murray, Marty Norden, Peter Rollins, Susan Rollins, John Tibbetts, and Ed Worteck. I have also talked to most of these individuals at length about Burns's television histories, as well as literally dozens of other helpful colleagues, most particularly Jean Baker, Ray Browne, Gary Burns, Rod Carveth, Bill Jones, Garth Jowett, Peter Lev, Terry Lindvall, Ray Merlock, John O'Connor, George Plasketes, Bob Toplin, and Jim Welsh, among many others too numerous to list but whose conversations with me were all greatly appreciated and essential in shaping my own developing ideas on the topic.

I want to finally acknowledge the support of Jo Ann Gora and Karen Gould at Old Dominion University for enabling me to take a sabbatical leave during which I wrote the majority of this manuscript. All of my colleagues in the Communication and Theatre Arts Department at Old Dominion have helped me immeasurably during the time when I was working on this book and I thank each and every one of them for their friendship and continuing support. My deepest gratitude goes to my family for their love and understanding. Being able to share parts of this project with Nan, Kate, and Mary Ellen as it took shape has made researching and writing *Ken Burns's America* an enjoyable experience from start to finish. They may not be as committed to and fascinated by documentaries as I am, but generously and with good humor they never let on. This book is dedicated to them.

CHAPTER ONE

Rebirth of a Nation: Reframing *The Civil War* (1990) on Prime-Time Television

A DEFINING MOMENT ON NATIONAL TV

Each generation, Lewis Mumford once said, rediscovers and reexamines that part of the past which brings the present new meaning and new possibilities.

—*Ken Burns, 1991*[1]

Ken Burns laughs now about the apprehension he felt on September 23, 1990, the day *The Civil War* premiered on prime-time television and changed his life forever. He had just completed a two-month promotional tour, a grueling process at which he is particularly adept, being a highly quotable and charismatic speaker and storyteller. He checked out of his midtown Manhattan hotel on that Sunday morning and began the long drive back to his home in Walpole, New Hampshire. Suddenly seized with misgivings, he remembers thinking long and hard about the remarks of several reviewers who predicted that *The Civil War* would be "eaten alive," going head-to-head with major network programming over five consecutive nights. That evening, he and his family were "completely unprepared for what was going to happen" next, as the first episode attracted 14 million viewers, while the full program reached nearly 40 million people by Thursday, the largest audience for a public television series ever. As Burns reminisced during a February 1993 interview, "I was flabbergasted! I still sort of pinch myself about it. It's one of the rare instances in which something helped stitch the country together, however briefly, and the fact that I had a part in that is just tremendously satisfying."[2]

So much about Ken Burns's career defies the conventional wisdom. He became one of public television's busiest and most celebrated producers during the 1980s, a decade when the historical documentary held little interest for most American TV viewers. He operates his own independent company, Florentine Films, in a small New England village more than four hours north of New York City, hardly a crossroads in the highly competitive and often insular world of corporately funded, PBS (Public Broadcasting Service) sponsored productions. His 16 major specials so far—*Brooklyn Bridge* (PBS, 1982), *The Shakers: Hands to Work, Hearts to God* (PBS, 1985), *The Statue of Liberty* (PBS, 1985), *Huey Long* (PBS, 1986), *Thomas Hart Benton* (PBS, 1989), *The Congress* (PBS, 1989), *The Civil War* (PBS, 1990), *Empire of the Air: The Men Who Made Radio* (PBS, 1992), *Baseball* (PBS, 1994), *The West* (PBS, 1996), *Thomas Jefferson* (PBS, 1997), *Lewis & Clark: The Journey of the Corps of Discovery* (PBS, 1997), *Frank Lloyd Wright* (PBS, 1998), *Not For Ourselves Alone: The Story of Elizabeth Cady Stanton and Susan B. Anthony* (PBS, 1999), *Jazz* (PBS, 2001), and *Mark Twain* (PBS, 2002)—are also strikingly out of step with the special effects and frenetic pacing of most nonfiction television, relying mainly on filmic techniques that were introduced literally decades ago. Most remarkably, however, 70 million Americans have now seen *The Civil War*. Fifty million have watched *Baseball;* and all of his other TV productions over the last decade have averaged an estimated 15 million viewers during their debut telecasts.[3] The cumulative popularity of Burns's biographical or quasi-biographical histories is striking by virtually any measure, and they have over time redefined the place of documentaries on prime-time television.

There are over 100 broadcast and cable networks in the United States today, and roughly 90 percent of these services have resulted from the dramatic rise of cable and satellite TV over the last 25 years. Scores of these cable networks have, moreover, become closely identified with documentaries in general and historical documentaries in particular. "Seven years after Ken Burns's *The Civil War* proved that history on TV could be engaging—and attract millions of viewers," announced *TV Guide* in 1997, "documentaries are all over the dial."[4] Burns has emerged as the signature figure for this far larger programming trend, primarily because of the unprecedented success of *The Civil War* as well as the consistently robust showings of his other television specials. Burns likewise has become a lightning rod for professional historians to express a spectrum of pro and con reactions about the growing popularity (especially with the general public) of films and television programs about the past overshadowing the one-time preeminence of written histories alone.

Television studies and biography as a branch of history, in retrospect, have long shared company as second-class citizens in academic life. It goes without saying that television has only recently emerged as a focus of serious inquiry within universities and colleges; one might even say that it has become a fashionable subject in a number of the humanities and social sciences. Biography, too, is making something of a comeback, although biographical scholarship certainly never resided as far out on the margins of academe as television studies.

The biographical approach probably reached its nadir in historical circles with the growing influence of the new social historians of the late 1960s and 1970s. This scholarly movement infused techniques mainly associated with the social and behavioral sciences into professional history, including a wide range of quantitative methodologies that succeeded in more effectively delineating the social, economic, and demographic aspects of their subjects. The old-style historical biographies appeared hopelessly unscientific and impressionistic in comparison, with their traditional reliance on narrative and their larger-than-life looks at "Great Men."

The most prominent and successful practitioners of the biographical approach to history during this era actually came from outside the academic world, led by best-selling writers, such as Shelby Foote with his three-volume, *Civil War: A Narrative* (1958-1974), David McCullough with early works such as *The Great Bridge* (1972), and Michael Shaara with his Pulitzer Prize–winning Civil War novel, *The Killer Angels* (1974).[5] Foote, McCullough, and Shaara among others were working within and renewing a much longer tradition of popular history, while also inspiring an even younger generation of nascent filmmakers who would initiate a minirevival in the historical documentary on television just a decade later. Ken Burns, in particular, adapted McCullough's *The Great Bridge* as his first film, *Brooklyn Bridge*, in 1982, and decided to produce *The Civil War* after he finished reading *The Killer Angels* on Christmas Day 1984, an experience he describes as "chang[ing his] life."[6] Burns further explains, "For nearly two centuries, we were animated entirely by the book, and what we knew about our past came from books, and we're going to need to restate the old heroes . . . [and] the old dramas, and we'll have to do it in a new visual way. And that's what I'm trying to do."[7]

The Civil War, in retrospect, is the ideal place to begin considering Burns as a highly influential television producer-director and popular historian who uses his work to explore the nation's historical legacy, reconfirming it in tandem with an audience of millions from a new generational perspective. This miniseries is both the watershed event in Burns's career as well as the impetus for the recent renaissance in documentary programming on American TV. All of Burns's narrative stylistics come together on a grand scale in *The Civil War*, as do his producing and promotional talents in shepherding a $3.2 million project from initial conception through final release.[8] Burns, additionally, emerges as a national celebrity as a result of the widespread public reaction and attention, which helps his subsequent television career but noticeably strains his relations with the historical community. *The Civil War* asserts in one fell swoop that history is no longer the principal domain of specialists, as it had been for nearly a half-century, but now is relevant and compelling for everyone—only this time on TV.

STRIKING A RESPONSIVE CHORD

I don't think [the story of the Civil War] can be told too often. I think surely
it ought to be retold for every generation.

—Ken Burns, 1990[9]

It's been more than a decade since the phenomenon of *The Civil War* premiered over five consecutive evenings (September 23-27, 1990), amassing a level of attention unsurpassed in public television history. Ken Burns's 11-hour version of the war acted as a kind of flash point for a new generation, attracting a spectrum of opinion that ranged from rapturous enthusiasm to milder interest in most segments of the viewing public, to outrage over Yankee propaganda in a few scattered areas of the South, to both praise and criticism from the academy.[10] Burns employed 24 consultants on this project including many prominent historians, but understandably not all of these scholars and filmmaking specialists agreed with everything in the final series.[11] With so many experts, and with a subject the size and scope of the Civil War as the historical terrain, a certain amount of controversy was unavoidable.[12]

One historian even concluded his analysis of *The Civil War* by calling the series "a flawed masterpiece," thus evoking the customary judgment of D. W. Griffith's *Birth of a Nation* (1915) that's been repeated in literally dozens of general film histories over the past 70 years.[13] This analogy only goes so far, however, making more sense on the grounds of shared cinematic brilliance than because of any similarities in outlook and sensibility. Indeed, one of Burns's stated intentions was to amend the "pernicious myths about the Civil War from *Birth of a Nation* to *Gone with the Wind*," especially in regard to racial stereotyping and the many other bigoted distortions in plot and imagery.[14]

Still, *Birth of a Nation* and *The Civil War* were similarly indicative of mainstream public opinion during their respective eras. For example, Russell Merritt has argued convincingly that the racist aspects of *Birth of a Nation* were anything but the ravings of some "isolated crackpot," but rather representative of white America at the time. According to Merritt, Griffith "attracted his audience . . . because the drama itself was one . . . Americans wanted to see."[15] As a result, *Birth of a Nation* was embraced by an estimated 10 percent of the U.S. population in its original release, making it the preeminent box office success in silent film history.[16]

The popular reaction to *The Civil War* was likewise lavish and record setting. Public television achieved its highest ratings ever when 38.9 million Americans tuned into at least one episode of the five-night telecast, averaging 12 million viewers at any given moment.[17] The audience research findings also indicated that half the viewership would not have been watching television at all if it had not been for this program.[18] This tendency was, moreover, reflected in the range of published responses to *The Civil War,* even including political pundits who rarely, if ever, attend to the opening of a major motion picture or television series. George Will, for example, wrote: "Our *Iliad* has found its Homer . . . if better use has ever been made of television, I have not seen it."[19] David Broder and Haynes Johnson weighed in with similar kinds of high praise.[20]

Film and television critics from across the country were equally effusive. *Newsweek* reported "a documentary masterpiece"; *Time* "eloquen[t] . . . a pensive epic"; and *U.S. News & World Report* "the best Civil War film ever made."[21] David Thomson in *Film Comment* declared that *The Civil War* "is the great American movie of the year—and one of the true epics ever made."[22] Tom Shales of the *Washington Post* remarked: "This is not just good television, nor even just great television. This is heroic television."[23] And Monica Collins of the *Boston Herald* informed her readers that "to watch 'The Civil War' in its entirety is a rare and wonderful privilege." She then urged: "You have to keep in mind that the investment in the program is an investment in yourself, in your knowledge of your country and its history."[24]

Between 1990 and 1992, accolades for Ken Burns and the series took on institutional proportions, as it garnered more than 40 major awards from the entertainment industry and the academic community combined. He won Producer of the Year from the Producers Guild of America; two Emmys for Outstanding Informational Series and Outstanding Writing Achievement; Best Foreign Television Award from the British Academy of Film and Television

Arts; a Peabody; a duPont-Columbia Award; a Golden Globe; a D. W. Griffith Award; two Grammys for Best Traditional Folk Album and Best Spoken Word; Best Special and Best Program from the Television Critics Association; and a People's Choice Award for Best Television Mini-Series. Gettysburg College also awarded *The Civil War* its first annual $50,000 Lincoln Prize as the "finest scholarly work in English on Abraham Lincoln or the American Civil War soldier" in competition with 41 books.[25] Burns, too, was given eight honorary doctorates from various American colleges and universities in 1991 alone.[26] As Burns remembers, "I don't really know how to put my finger on it. A generation ago as we celebrated, or tried to celebrate the centennial, we seemed focused on the battles or the generals, and the kind of stuff of war, but here we seemed to respond to the human drama and maybe it just resonated in a particular way with how we are. I feel a tremendous sympathy for this country and somewhere along the line that sympathy must line up with where we are now and whatever the subject is."[27]

The Civil War became a phenomenon of popular culture. The series was mentioned on episodes of *Twin Peaks, Thirtysomething,* and *Saturday Night Live* during the 1990-1991 television season. It was spoofed on National Public Radio and in a *New Yorker* cartoon. Ken Burns appeared on the *Tonight Show* shortly after Johnny Carson took the unusual step of recommending the series to his audience on the Monday following the Sunday debut of the first episode. And Burns was selected by the editors of *People* magazine as one of their "25 most intriguing people of 1990," along with their usual odd assortment of international celebrities, including George Bush, Julia Roberts, M. C. Hammer, Saddam Hussein, Bart Simpson, Sinead O'Connor, and Nelson Mandela.[28]

The series also developed into a marketing sensation as the companion book by Knopf, *The Civil War: An Illustrated History,* became a runaway bestseller. According to *Publishers Weekly,* "The celebrated PBS television series *The Civil War* certainly helped its eponymous companion volume sell enough books for the #2 slot. Knopf reported sales of 560,931 in 1990, and the book is still enjoying a brisk rate of sales in 1991."[29] This hardcover title spent 11 straight weeks on the top-ten list during 1990, and then extended this streak for 15 additional weeks in 1991.[30] "Considering the $50 ticket price," the industry trade publication related, "the book is easily the year's bestselling nonfiction grosser in dollars."[31] The accompanying Warner soundtrack and the nine-episode videotaped version from Time-Life were similarly successful. Burns noted, for example, that "The Civil War videotapes are the best-selling nonfiction documentary series on history ever made,"[32] as

Billboard cited that it reached the one-million plateau in aggregate sales as early as October 1993.[33]

Several interlocking factors evidently contributed to the extraordinary level of interest surrounding *The Civil War,* including the overall technical and dramatic quality of the miniseries itself, its accompanying promotional campaign, the momentum of scheduling Sunday through Thursday, the synergetic merchandising of all its ancillary products, and a TV industry strike earlier in the year which disrupted the fall season and caused the network competition to briefly delay its season premieres. Most significantly, though, a new generation of historians had already begun addressing the war from the so-called bottom-up perspective, underscoring the role of African Americans, women, immigrants, workers, farmers, and common soldiers in the conflict. This fresh emphasis on social and cultural history had revitalized the Civil War as a subject, adding a more inclusive and human dimension to the traditional preoccupations with great men, transcendent ideals, and battle strategies and statistics. The time was again propitious for creating another rebirth of the nation on film which included the accessibility of the bottom-up approach. In Ken Burns's own words, "I realized the power that the war still exerted over us."[34]

The Civil War has, indeed, fascinated Americans for more than 140 years. James M. McPherson, the 1988 Pulitzer Prize–winning author of *Battle Cry of Freedom* estimates that the literature "on the war years alone . . . totals more than 50,000 books and pamphlets."[35] Reader interest had actually been increasing in the five years preceding the debut of *The Civil War;* 520 of the 1,450 titles that were still in print in September 1990 had only been published since 1986. After the premiere of the series, however, fascination with the war became "higher . . . than it has ever been."[36]

Shelby Foote was actually the first contemporary writer to liken the Civil War to the *Iliad* in the third volume of his trilogy, and his intent was to emphasize how "we draw on it for our notion of ourselves, and our artists draw on it for the depiction of us in the same way that Homer and the later dramatists—Aeschylus, Sophocles, Euripides—drew on the Trojan war for their plays."[37] Much of the success of Ken Burns's *The Civil War* must be equated in kind to the extent that his version made this nineteenth-century conflict immediate and comprehensible to audiences in the 1990s. The great questions of race and continuing discrimination, of the changing roles of women and men in society, of big government versus local control, and of the individual struggle for meaning and conviction in modern life all remain. The Civil War captivates because its purposes endure; Americans are as engaged as ever in the war's dramatic conflicts. As Burns summarizes, there is so much about *The Civil War* that reverberates today . . . a developing women's movement, Wall Street speculators, the imperial presidency, new military technology, the civil rights question and the contribution of black soldiers . . . there are also approximations and that sort of thing. You have to cut stuff out. I would have loved more on the congressional sort of intrigues during the Civil War. I would have loved to do more on women and more on emancipation and more on Robert E. Lee and more on the western battles, but limitations of photographs or just time or rhythm or pacing, or whatever it is, conspired against those things. And they were there, but they were taken out to serve the demands of the ultimate master, which is narrative.[38]

THE CIVIL WAR AS MADE-FOR-TELEVISION HISTORY

This is our great epic poem. This is this country's great narrative, like the *Mahabarata* or Homer's *Iliad*.

—*Ken Burns, 1990*[39]

Narrative is a particular mode of knowledge and means of relaying history. It is a historical style that is dramatic and commonly literary, although *The Civil War* does indicate that it can be ideally adapted to the electronic media as well. Ken Burns strongly recognizes that "television has become more and more the way we are connected to the making of history."[40] In selecting the Homeric model, he is choosing certain narrative parameters that are epic and heroic in scope. The epic form tends to celebrate a people's national tradition in sweeping terms; and a recurring assertion throughout Burns's filmic history is how the Civil War gave birth to a newly redefined American nation. The final episode, "The Better Angels of Our Nature," for example, begins with three commentaries on nationhood that rhetorically set the stage from which the series will be brought to its rousing conclusion:

> Strange is it not that battles, martyrs, blood, even assassination should so condense a nationality.
>
> —*Walt Whitman (as spoken by Garrison Keillor)*

> It is *the* event [the Civil War] in American history in that it is the moment that made the United States as a nation.
>
> —*Barbara Fields*

Before the war it was said the United States *are*, grammatically it was spoken that way and thought of as a collection of independent states, and after the war it was always the United States *is* as we say today without being self-conscious at all—and that sums up what the war accomplished: it made *us* an *is*.

—Shelby Foote

These remarks are then immediately followed by the bittersweet and tragic lament that serves as the series' anthem, "Ashokan Farewell," thus reinforcing the overall heroic dimensions of the narrative. Heroism, honor, and nobility are related Homeric impulses that permeate this series, shaping our reactions to the great men of the war, such as Abraham Lincoln, Frederick Douglass, and Robert E. Lee, along with the many foot soldiers whose bravery often exceeded the ability of their officers to lead them, resulting in the appalling carnage recounted in episode after episode.

History on TV tends to stress the twin dictates of narrative and biography which ideally expresses television's inveterate tendency toward personalizing all social, cultural, and for our purposes, historical matters within the highly controlled and viewer-involving confines of a well-constructed plot structure. The scholarly literature on television has established intimacy and immediacy (among other aesthetics) as inherent properties of the medium.[41] In the case of intimacy, for instance, the confines of the relatively small TV screen which is typically watched within the privacy of the home environment have long resulted in an evident preference for intimate shot types (i.e., primarily close-ups and medium shots), fashioning most fictional and nonfictional historical portrayals in the style of personal dramas or melodramas played out between a manageable number of protagonists and antagonists. When successful, audiences closely identify with the historical actors and stories being presented, and, likewise, respond in intimate ways in the privacy of their own homes.

The series' most celebrated set piece, the poignant and eloquent voice-over of Major Sullivan Ballou's parting letter to his wife before he was killed at the first battle of Bull Run (again accompanied by the haunting strains of "Ashokan Farewell"), illustrates the skillful way in which Burns regularly infuses the epic sweep of *The Civil War* with a string of highly personal and well-placed dramatic interludes. This scene, which lasts approximately three and one-half minutes, concludes episode one, "1861—The Cause," thus imbuing the preceding 95 minutes an air of melancholy, romance, and higher purpose. Poetic license is used throughout the segment, as Ballou's declaration of love is heard over images that have nothing factually to do with Sullivan Ballou but evoke the emotional texture of his parting sentiments: photographs of the interior of a tent where such a letter might have been written, a sequence of pictures portraying six other Civil War couples, and three static filmed shots of Manassas battlefield as it looks today in a pinkish twilight.

After narrator David McCullough briefly begins the scene with the words, "a week before Manassas, Major Sullivan Ballou of the 2nd Rhode Island wrote home to his wife in Smithfield," actor Paul Roebling's serenely heartfelt and understated reading fades up quietly behind the photographs:

> My very dear Sarah . . . I feel impelled to write a few lines that may fall under your eyes when I shall be no more . . . I have no misgivings about, or lack of confidence in the cause in which I am engaged, and my courage does not halt or falter . . . Sarah my love for you is deathless . . . the memories of the blissful moments I have spent with you come creeping over me . . . but O Sarah, if the dead can come back to this earth and flit unseen around those they loved, I shall always be near you . . . always, always, and if there be a soft breeze upon your cheek, it shall be my breath, as the cool air fans your throbbing temple, it shall be my spirit passing by. Sarah do not mourn me dead; think I am gone and wait for thee, for we shall meet again.

The impact and effectiveness of this section, entitled "Honorable Manhood," was immediately apparent as Ken Burns recalled a year later: "Within minutes of the first night's broadcast, the phone began ringing off the hook with calls from across the country, eager to find out about Sullivan Ballou, anxious to learn the name of Jay Ungar's superb theme music ("Ashokan Farewell"), desperate to share their families' experience in the war or just kind enough to say thanks. The calls would not stop all week—and they continue still."[42]

Burns's plot structures are characteristically composed of four different kinds of scenes. To start with, he employs *narrative descriptions* which primarily move the story along. These sections follow a simple chronology and are designed above all else to provide the audience with the basic historical facts and figures on what is happening and who is involved in the action at any given time. *The Civil War* was originally planned as a five-part, five-hour series, according to the National Endowment for the Humanities (NEH) grant application written in late 1985 and early 1986 by Ken and Ric Burns, with each section

> covering roughly one year of the conflict, 1861 through 1865. While the three central episodes will treat most of the major battles and campaigns, we will take advantage of the militarily less eventful years, 1861 and 1865, to

explore the origins and consequences of the conflict. The war was, of course, a great epic, and episode by episode we will chart the large ebb and flow of the war: the mobilization of men and material, of industry and new technology, the deeds of generals and diplomats, the statistics of death, disease, and cost.[43]

By the time of the premiere telecast, *The Civil War* had more than doubled in size to 11 hours where Burns "eventually subdivided '62, '63, '64, and '65 into the first and second halves of the years, creating a total of nine episodes."[44] As Daniel Boorstin explains, the "most popular" method of organizing historical stories is in yearly, decade-long, and "hundred year packages. Historians like to bundle years in ways that make sense, provide continuity and link past to present."[45] Burns, first and foremost, then, creates narratively descriptive scenes that provide the factual details needed to support and validate the larger historical outlines of the overall nine-episode structure.

Second, he designs what he calls *emotional chapters,* such as the aforementioned Sullivan Ballou set piece, which have the "ability to float between episodes."[46] This category of scene is less bound by chronological demands than by its capacity to affect mood and engage an audience emotionally at strategic moments within the plot. *The Civil War,* for instance, is peppered throughout with the entertaining and informative anecdotes of writer, popular historian, and master raconteur Shelby Foote. His seemingly intimate asides about the human interest aspects of the conflict add a needed personal dimension to the drier evidential framework of the broader historical narrative. As Burns describes: "Just go back to the section on the Gettysburg Address and watch Shelby's head twitch as he talks about Lincoln stepping down from the stand. And Shelby says '[Lincoln] came back, and he turned to his friend Ward, and he said, "Ward, that speech won't scour."' And [Shelby] tilts his head, as if, [had] the camera pulled back, you'd see next to Shelby, Abraham Lincoln, and on the other side of Lincoln, Ward Lamon. And to me, any man who puts you there, that's a great gift."[47]

The third type of scene that Burns designs are those he calls "*telegrams,* [or] short bursts that also have a certain potential to move but are more or less tied to a specific moment or a specific time."[48] Telegrams, in a sense, are a mixture of both the scenes of narrative description, because they are bound to whatever event is transpiring in the storyline at the time, and the emotional chapters since these concise segments strongly contribute to viewer involvement. Prime examples of this sort of scene include the many private reactions to a wide array of historical developments throughout the series by either southern diarist Mary

Chestnut (as spoken by Julie Harris) or northern lawyer and civic leader George Templeton Strong (as spoken by George Plimpton). The most remembered telegrams, undoubtedly, are the ones built around single archival photographs featuring ground level views of ordinary Union and Confederate soldiers before and after virtually every bloody engagement. These evocative images, once again, render the personal dimension of the conflict that much more accessible and identifiable to a contemporary audience of millions. As Burns discloses,

> we wanted you to believe you were there . . . there is not one shot, not one photograph of a battle ever taken during the Civil War. There is not one moment in which a photographer exposed a frame during a battle, and yet you will swear that you saw battle photography . . . You live inside those photographs, experiencing a world as if it was real inside those photographs . . . Once you've taken the poetry of words and added to it a poetry of imagery and a poetry of music and a poetry of sound, I think you begin to approximate the notion that the real war could actually get someplace, that you could bring it back alive.[49]

Ken Burns, fourth and finally, constructs *editing clusters* as his way of critically analyzing the various sides of a theme, question, or controversy that is central to a better overall understanding of his subject, such as "slavery and emancipation" in *The Civil War*, which he calls "the inner core of our story."[50] This type of scene involves cutting together images of historical relevance with a montage of commentators who typically present both corroborating and conflicting opinions, creating a collage of multiple viewpoints. In the "Was It Not Real?" segment of the final episode entitled "The Better Angels of Our Nature," for example, there is a montage of three commentaries presenting both confirming and dissenting points of view about the lasting meaning of the Civil War. Barbara Fields, who previously had suggested that Lincoln was actually a moderate on the issue of race in comparison to his contemporaries, begins by observing that "the *slaves won the war* [my emphasis] and they lost the war because they won their freedom, that is the removal of slavery, but they did not win freedom as they understood freedom."

Next, James Symington provides a different slant on the issue by declaring that "the significance of *Lincoln's life and victory* [my emphasis] is that we will never again enshrine [slavery] into law," while affirming Fields with "let's see what we can do to erase . . . the deeper rift between people based on race . . . from the hearts and minds of people." Stephen Oates then ends this section by shifting the focus to the survival and triumph of "popular government," ending with the assertion that the Civil War is "a testament to the liberation of the human spirit for all time." Oates's conclusion has little to do with the specific substance addressed in the previous statements by either Fields or Symington, although coming where it does, his testimony cannot help but soften the references to racial injustice that preceded it.

More importantly, this specific editing cluster establishes the liberal pluralist consensus: in other words, different speakers might clash on certain issues (such as what degree of freedom was actually won in the Civil War and by whom), but disagreements ultimately take place within a broader framework of agreement on underlying principle. In this case, the larger principle is Oates's evocation of popular government, which is understood to guarantee the democracy and human rights needed to eventually eradicate racial inequality and disharmony. A historical narrative, therefore, does not merely record and dramatize what happens; it also at times interprets events and shapes the presentation of the subject at hand.

Furthermore, this particular example illustrates that the historical documentary is able to sustain a certain degree of analysis (although not nearly as deeply and comprehensively as written discourse or public discussion and debate). The expert testimonies and first-person reports that Burns employs do provide shifting angles of vision that sometimes agree and, at other times, differ and contrast with each other. These multiple voices, however, form a cultural consensus because of both the filmmaker's liberal pluralist orientation and, in Burns's words, "the power of film to digest and synthesize."[51] In the end, then, Ken Burns, the popular historian, is much more a committed storyteller than a reasoned and detached analyst. As he explains, "it is the texture of emotion that is important to me. And this is what television can do that all the texts cannot do."[52]

Burns's position as a historical documentarist, moreover, straddles two well-established and generally distinct professions. He is a highly accomplished television producer-director, and as he often characterizes himself, "an amateur historian" with a wide-ranging interest in American history but no special scholarly training or specialization in any one particular area. His work habits, nevertheless, do have a great deal in common with many standard academic practices. Preparing each historical documentary includes the disciplined rigors of thoroughly researching his subject, writing grant proposals, collaborating and debating with an assortment of scholarly advisors, composing multiple drafts of the offscreen narration, as well as gathering and selecting the background readings and the expert commentaries. (The final 372-page script for *The Civil War*, for instance, was its fifteenth version.)[53] The academic community also began paying far closer critical attention to Burns and his made-for-television histories, particularly after the remarkable public response to this miniseries.

One historian, for example, chided Burns for utilizing the Sullivan Ballou letter without "report[ing] in *The Civil War* . . . that the letter was never sent; it was discovered among Ballou's possessions."[54] Another scholar raised the question that a number of different versions of the letter do, in fact, exist.[55] Burns responds that "poetic license is that razor's edge between fraud and art that we ride all the time. You have to shorten, you have to take shortcuts, you have to abbreviate, you have to sort of make do with, you have to sometimes go with something that's less critically truthful imagery-wise because it does an ultimately better job of telling the larger truth, but who is deciding and under what system becomes the operative question."[56]

Here Burns raises two fundamental differences between his own approach to producing history on television and the academic standards shared by most professional historians. First, he is far more concerned with the art of storytelling than with maintaining detailed accuracy in a fundamentalist sense, although he is always careful to marshal the facts of history as much as his stated goal of capturing the emotional truth of his subject allows.[57] As he explains, "the historical documentary filmmaker's vocation is not precisely the same as the historian's, although it shares many of the aims and much of the spirit of the latter . . . The historical documentary is often more immediate and more emotional than history proper because of its continual joy in making the past present through visual and verbal documents."[58] Second, Ken Burns is not as self-reflexive about historiography as are professional historians. He is aware that there are "systems" to history, but there are times when he is criticized for stressing plot over historical analysis:

> I am primarily a filmmaker. That's my job. I'm an amateur historian at best, but more than anything if you wanted to find a hybridization of those two professions, then I find myself an emotional archeologist. That is to say, there is something in the process of filmmaking that I do in the excavation of these events in the past that provoke a kind of emotion and a sympathy that remind us, for example, of why we agree against all odds as a people to cohere.[59]

At first blush, this final statement might appear to confirm the assessment offered in a 1992 *American Quarterly* essay, "Videobites: Ken Burns's 'The Civil War' in the Classroom," which suggests that "'The Civil War' stands as a new nationalist synthesis that in aims and vision can be most instructively compared to James Ford Rhodes's histories of the Civil War (written at the end of the nineteenth and in the early twentieth centuries)."[60] A 1991 appraisal in *American*

Historical Review similarly takes the filmmaker to task: "Burns used modern historical techniques, at the level of detail and anecdote, to create an accessible, human-scale account of the Civil War. But, when it comes to historical interpretation, to the process by which details coalesce to make events meaningful, *The Civil War* is vintage nineteenth century." The severity of these judgments are encapsulated by the same author in a final dismissal: "[*The Civil War*] is the visual version of the approach taken by generations of Civil War buffs, for whom reenacting battles is a beloved hobby."[61]

Historical documentaries should certainly be subject to evaluation and criticism, especially if they are to be viewed by audiences of tens of millions on TV or in theaters and subsequently used as teaching tools in our nation's schools. *The Civil War,* for example, was licensed after its premiere telecast to over 60 colleges and universities for future classroom use, and Ken Burns reports that he's "received over 6,000 letters and cards from secondary school teachers alone, grateful for the series, pleased with how well it works."[62] There is clearly a responsibility to assess any film being employed for educational purposes to such a widespread degree. Both of these articles, in fact, do raise important questions of interpretation and detail that are useful and edifying. It is a welcome development that historians are increasingly attending to the validity of films and television programs.

These reviews, on the other hand, concurrently demonstrate the academy's longstanding and persistent tendency to underestimate yet another motion picture or television series which, in turn, shortchanges *The Civil War* as popular history. One of the primary goals of scholarship is to create new knowledge and be cutting edge. No more thorough indictment exists, according to this frame of reference, than to reject a text for its obsolete conception and design; in this case, banishing it to the nineteenth century. *The Civil War,* however, deserves a more measured examination than merely being dismissed as the stuff of "Civil War buffs."

In his widely acclaimed book *That Noble Dream* (1988), Peter Novick has skillfully examined the controversies that have fundamentally affected the historical discipline over the last generation.[63] Current debates continue in the literature and at conferences over the relative merits of narrative versus analytic history, synthetic versus fragmentary history, and consensus versus multicultural history. Lawrence Levine suggests that this series of historiographical exchanges makes "sense only when it is seen as what, at its root, it really is—a debate about the extent to which we should widen our historical net to include the powerless as well as the powerful, the followers as well as the leaders, the margins as well as the center, popular and folk culture as well as high culture."[64]

The Civil War is a product of this intellectual climate. In this respect, it is not enough to focus on specific details from *The Civil War,* such as the Sullivan Ballou letter, without also considering Ken Burns's ideological bearings alongside the scope of his historical net. This more comprehensive outlook reveals fragments of a nationalist approach to historiography as the aforementioned reviewers suggest. *The Civil War* also evinces elements of the romantic, progressive, social history, and consensus schools as well. As Burns explains, "in narrative history you have this opportunity, I believe, to contain the multitude of perspectives. You can have the stylistic, and certainly my films have a particular and very well known style. You can involve yourself with politics, but that's not all there is. And that's what I'm trying to do, is to embrace something that has a variety of viewpoints."[65]

The Civil War is essentially a pastiche of assumptions derived from a number of schools of historical interpretation. As just mentioned, the series is nationalistic in its apparent pride in nation building, but without the nineteenth-century arrogance that envisioned America as the fulfillment of human destiny. *The Civil War* is romantic in its chronological and quasi-biographical narrative, but it lacks the unqualified, larger-than-life depictions of the unvarnished great men approach. *The Civil War* is progressive in its persistent intimation that the war was ultimately a struggle to end slavery and ensure social justice, although this perspective, too, is tempered by some passages, such as Barbara Fields's assertion in the final episode that the Civil War "is still to be fought, and regrettably, it can still be lost."

The Civil War is also informed by social history with its attention to African Americans, women, laborers, farmers, and especially firsthand accounts in each of the nine episodes by two common soldiers (i.e., Elisha Hunt Rhodes, a Yankee from Rhode Island, and Sam Watkins, a Confederate from Tennessee), but the series is nowhere near as representative of the bottom-up view as is pure social history. In Burns's own words, "I try to engage, on literally dozens of levels, ordinary human beings from across the country—male and female, black and white, young and old, rich and poor, inarticulate and articulate."[66]

What Ken Burns is annunciating is his liberal pluralist perspective where differences of ethnicity, race, class, and gender are kept in a comparatively stable and negotiated consensus within the body politic. Burns's brand of made-for-television history is marked more by agreement than is the multicultural or diversity model that grounds the social history perspective. The preservation of the Union and an emphasis on its ideals and its achievements are fundamental to consensus thinking; they are also some of Burns's primary themes throughout *The Civil War:*

It's interesting that we Americans who are not united by religion, or patriarchy, or even common language, or even a geography that's relatively similar, we have agreed because we hold a few pieces of paper and a few sacred words together, we have agreed to cohere, and for more than 200 years it's worked and that special alchemy is something I'm interested in. It doesn't work in a Pollyanna-ish way . . . we corrupt as much as we construct, but nevertheless, I think that in the aggregate the American experience is a wonderful beacon . . . and I think the overwhelming response to *The Civil War* is a testament to that.[67]

Rather than being ideologically stuck in the nineteenth century, Ken Burns and the audience for *The Civil War* were instead fully contemporary in their outlook. The tenets of liberal pluralism were prevalent throughout American culture during the 1990s and continue to be today. Popular metaphors, such as the quilt or the rainbow or, to a lesser degree, the old-fashioned melting pot, are still widespread images used by public figures across the political spectrum to evoke a projection of America that is basically fixed on agreement and unity, despite whatever social differences may exist. By realizing this perspective on film, Ken Burns has, moreover, usurped one of the foremost goals of social history, which is to make history meaningful and relevant to the general public. *The Civil War* brilliantly fulfills this objective as few books, or motion pictures, or television series, or even teachers for that matter, have ever done before.

BRIDGING THE DIVIDE
BETWEEN POPULAR AND PROFESSIONAL HISTORY

We have begun to use new media and new forms of expression—including films and television—to tell our histories, breaking the stranglehold the academicians exercised over this discipline for the last hundred years.

—*Ken Burns, 1991*[68]

My job is to convey history to people. No film, however well done, can ever replace that task.

—*Barbara Fields, 1990*[69]

The mutual skepticism that sometimes surfaces between popular and professional historians is understandable and unfortunate. Each usually works with different media (although some scholars do produce historical TV programs, videos, and films); each tends to place a dissimilar stress on the respective roles of storytelling versus analysis in relaying history; and each tailors a version of the past that is designed for disparate—though overlapping—kinds of audiences. These distinctions are real enough. Still, the artist and the scholar, the amateur and the expert can complement each other more than is sometimes evident as they both make their own unique contributions to America's collective memory, a term referring to the full sweep of historical consciousness, understanding, and expression that a culture has to offer.

Interdisciplinary work in memory studies now boasts adherents in American studies, anthropology, communication, cultural studies, English, history, psychology, and sociology.[70] The contemporary preoccupation with memory dates back to Freud, although recent scholarship focuses more on the shared, collective nature of remembering rather than the individual act of recalling the past, which is the traditional realm of psychological inquiry into this topic. Researchers today, most importantly, make distinctions between the academic historical record and the rest of what is referred to as collective memory. Professional historians, in particular, "have traditionally been concerned above all else with the accuracy of a memory, with how correctly it describes what actually occurred at some point in the past."[71] "Less traditional historians have [recently] allowed for a more complex relationship, arguing that history and collective memory can be complementary, identical, oppositional, or antithetical at different times."[72] According to this way of thinking, more popular uses of memory have less to do with accuracy per se than with using the past as a kind of communal, mythic response to current events, issues, and challenges. The proponents of memory studies, therefore, are most concerned with how and why a remembered version is being constructed at a particular time, such as *The Civil War* in 1990, than with whether a specific rendition of the past is historically correct and reliable above all else. As Ken Burns further clarifies his approach, "history . . . is an inclusion of myth as well as fact because myth tells you much more than fact about a people."[73]

Rather than think of popular and professional history as diametrically opposed traditions (i.e., one unsophisticated and false, the other more reliable and true), it is more helpful, instead, to consider them as two ends of the same continuum. In his 1984 book *Culture as History,* the late Warren Susman first championed this more sympathetic appreciation of the popular historical tradition. Susman noted that myth and history are intimately linked to each other. One supplies the drama; the other the understanding. The popular heritage holds the potential to connect people passionately to their pasts; the scholarly camp maps out the processes for comprehending what actually happened with richness and depth. Susman's fundamental premise was that popular history and professional history need not always clash at cross-purposes.

From this more inclusive perspective, popular history and professional history are seen less as discrete traditions, and more as overlapping parts of the same whole, despite the many tensions that still persist. For instance, popular histories can nowadays be recognized for their analytical insights, while professional histories can similarly be valued for their expressive possibilities. Popular history, too, is built squarely upon the foundations of academic scholarship; it provides professional historians, such as Barbara Fields and Stephen Oates in the case of *The Civil War,* with a much-higher-profile platform from which to introduce their scholarly ideas and insights to a vastly wider audience. Together they enrich the historical enterprise of a culture, and the strengths of one can serve to check the excesses of the other.[74]

Any understanding of *The Civil War,* accordingly, needs to be based on the fundamental assumption that television's representation of the past is an entirely new and different kind of history altogether. Unlike written discourse, the language of TV is highly stylized, elliptical (rather than linear) in structure, and associational or metaphoric in the ways that it portrays its historical themes, figures, and events. *The Civil War* as popular history is above all an artistic attempt to link audiences immediately and intensely with the life stories of the people who were caught up in the conflict. One historian, in fact, content analyzed "444 substantive letters from the more than 1,100 letters Burns had received as of March 1991," finding that "more than one out of every four letters (27 percent) praised Burns for offering them a sense of direct, emotional connection with the past."[75] Like any mediated rendering of history, *The Civil War*'s main strength is experiential, meaning that it provides the dramatic illusion for viewers that they are somehow personally involved in the action, even as they are learning factual details about this vast subject through the course of following the narrative. Popular history is always vicarious and participatory, rather than comparatively detached and analytical like most examples of written professional history.

Made-for-television histories are, thus, never conceived according to the standards of professional history. They are not intended chiefly to debate issues, challenge the conventional wisdom, and create new knowledge and perspectives. *The Civil War,* more specifically, is designed for the less contentious and communally oriented environment of prime-time television with its audiences in the tens of millions. In this way, the act of producing, telecasting, and viewing this miniseries became a large-scale cultural ritual in and of itself. This process, in turn, completed three important functions: First, *The Civil War* served as an intermediary site bridging the findings of professional historians with the interests of the general public. "There are levels of inquiry," according to Burns, "and we have to celebrate those that bring us to the door of the next level. And I think *Roots* brings in a huge audience. Maybe *The Civil War* has a little bit more select audience . . . but all of it is enriching the academy as well as the populace."[76] Second, the series facilitated an ongoing negotiation with America's usable past by examining those parts of the collective memory that were most relevant to the television producers as well as the nearly 40 million viewers who decided to tune in, issues of widespread public concern such as the residual effects of slavery and the continuance of racial conflict and discrimination in the United States, the influence of a strong federal presence in both state and local governments across the country, and the search for meaning and personal responsibility in national life. In this regard, "[h]istory [to Burns] is really not about the past; it's about the present. We define ourselves now by the subjects we choose from the past and the way each succeeding generation interprets those subjects. They are more a mirror of how we are now than they are a literal guide to what went before."[77] And third, *The Civil War* loosely affirmed majoritarian standards, values, and beliefs, or in the filmmaker's own words, "there is more *unum* than *pluribus* in my work."[78]

Ken Burns, overall, articulates a version of the country's past that conveys his own perspective as a popular historian, intermingling many widely held assumptions about the character of America and its liberal pluralist aspirations. Like other documentarians of his generation, he, too, addresses matters of race, gender, class, and regional division, but unlike many of his contemporaries, he presents an image of the United States eventually pulling together despite its many chronic differences rather than a society coming apart at the seams. Exploring the past is Burns's way of reassembling an imagined future from a fragmented present. *The Civil War,* in particular, reaffirmed for the members of its principal audience (which skewed white, male, 35 to 49, and upscale in the ratings) the relevance of their past in an era of unprecedented multicultural redefinition.[79] This aesthetic reintegration of the past into the present is one of the major purposes of popular history. For Ken Burns, it is a process of reevaluating the country's historical legacy and reconfirming it from a wholly new generational outlook.

KEN BURNS'S LIVING ROOM WAR

I began the project thinking that it would be an antiwar piece—and it is that— it's a Trojan Horse, it is our story. And the suffering that we inflicted on one another reminds us in a much clearer voice about the suffering that all wars inflict on human beings. I don't think that I would call myself a pacifist, though that end was certainly in mind. I found that the project became much more complex. There are just wars and I hope I would have had the good sense to fight in it for principles that I believe.

—*Ken Burns, 1990*[80]

This quotation is Ken Burns's answer to a journalist's question at an October 29, 1990 National Press Club newsmaker luncheon in Washington, D.C. about whether or not he considered himself a pacifist. His response suggests how he, as a television producer-director, drew upon the ideologies of the present as well as the mythic formulas of the past to create his own particular version of the Civil War. Certainly part of the pervading sensibility of this miniseries is rooted in the common experience of many Americans who had seen numerous images of graphic violence and suffering during America's first full-scale television war, in Vietnam (1962-1975). Burns, in many ways, is the ideal filmmaker for this period of transition between generations, bridging the beliefs and ideals of the people who came of age during World War II with his own frame of reference as a baby boomer. He agrees that his perspective was shaped by both "the fifties and sixties because I think that maybe all of that stimulus from the centennial celebration of the Civil War to the mythology that still pertained not only got fixed but then got challenged in the sixties. And I think that those two things going in opposite directions probably accounts for why we're all drawn to [*The Civil War*] right now."[81]

The debut telecast of *The Civil War*, moreover, occurred in the wake of the Persian Gulf conflict. Iraq had just invaded Kuwait, on August 2, 1990. An imminent escalation of hostilities was continually on the minds of most Americans.[82] Corresponding TV coverage of a massive military buildup in the region provided the immediate backdrop for the series premiere on September 23, just seven and one-half weeks later. Revisiting the country's quintessential war on television also meant reconsidering all of the essential themes of *The Civil War* that Burns and his colleagues had so eloquently captured. The filmmakers were eager to thoroughly reassess the American Civil War from their own generational viewpoint, reclaiming whatever relevance it still might hold for themselves personally as well as for future audiences of the miniseries.

The Civil War, like the *Iliad,* is a hereditary story of epic and tragic proportions. Burns even remembers Shelby Foote telling him during the early stages of production that "God is the greatest dramatist—just tell the story."[83] The seeming inevitability of events as depicted in *The Civil War* is, thus, given a near mythic presentation. Once pivotal moments appear on screen, such as the firing on Fort Sumter or the battle at Gettysburg, there is a driving momentum to the narrative which tacitly connotes that circumstances are now unstoppable (if not entirely preordained). Even the larger-than-life protagonists of the series—for example Grant, Sherman, Lee, and Jackson (as with the *Iliad*'s Achilles, Aias, Hector, and Diomedes), have their courage, skill, and honor fully tested by capable and worthy adversaries. Some figures, too, such as McClellan (as with Agamemnon), simply do not measure up to the heroic challenges and responsibilities that are thrust upon them. Lincoln, unlike anyone else, emerges as the remembered embodiment of *areté* (the Greek term for excellence of character); he is *The Civil War*'s most shining paragon who Burns even now refers to as "quite simply the most important man I've ever gotten to know. He has helped me know myself more clearly and helped define my own inscrutable love of country."[84] For the producer-director, then, Lincoln is the highest incarnation of the American spirit in the miniseries; his epic example is as relevant today in myth as his actual behavior was in national affairs 140 years ago.

As expected, Burns's contemporaneous treatment of myth in *The Civil War* is markedly more democratic in outlook and execution than Homer's classical handling of the Trojan War in the *Iliad*. *The Civil War* focuses on ordinary soldiers and civilians as well as the expected generals and politicians. The audience is, therefore, afforded a ground-level view of the historical actors through the sheer volume and variety of different people pictured in the more than 3,000 black-and-white archival photographs used throughout the series. These images regularly spotlight the facial expressions of young combatants or runaway slaves or nurses tending to the wounded as each of these bottom-up participants engages the viewer with looks that command recognition and reciprocal involvement. The narration, too, is delivered compellingly with resonant matter-of-factness by David McCullough, emphasizing once again the human dimension of the conflict with an almost Whitmanesque enthusiasm for naming names and chronicling the experiences of everyday Americans. From episode one, "1861—The Cause," for example,

> The Civil War was fought in 10,000 places, from Valverde, New Mexico, and Tullahoma, Tennessee, to St. Albans, Vermont, and Fernandina on the Florida coast. More than 3 million Americans fought in it, and over 600,000 men, 2 percent of the population, died in it . . . In two days at Shiloh, on the banks of the Tennessee River, more Americans fell than in all previous American wars combined. At Cold Harbor, some 7,000 Americans fell in twenty minutes. Men who had never strayed twenty miles from their front doors now found themselves soldiers in great armies fighting epic battles hundreds of miles from

home. They knew they were making history, and it was the greatest adventure of their lives.

Despite all these historical details, the filmmakers also imbue *The Civil War* with a mythic ambience by generously utilizing the many human interest aspects of the subject. Within the first 30 seconds of the series, for instance, a cyclical structure common to many traditional myths is suggested for the 11-hour narrative with the stranger-than-fiction story of Wilmer McLean, a folksy coincidence which only rings true because it is so implausibly down to earth:

> By the summer of 1861, Wilmer McLean had had enough. Two great armies
> were converging on his farm, and what would be the first major battle of the
> Civil War—Bull Run or Manassas as the Confederates called it—would soon
> rage across the aging Virginian's farm, a Union shell going so far as to tear
> through his summer kitchen. Now McLean moved his family away from
> Manassas, far south and west of Richmond—out of harm's way he prayed—
> to a dusty little crossroads town called Appomattox Court House. And it was
> there in his living room three and a half years later that Lee surrendered to
> Grant, and Wilmer McLean could rightfully say, "The war began in my front
> yard and ended in my front parlor."

The final nine-minute section of the last episode entitled "Was It Not Real?" contains yet another revealing personal touch which further humanizes the four-year ordeal. In this emotional chapter, Sergeant Barry Benson, a South Carolina veteran, is featured during the closing 2 minutes and 45 seconds of the series in a concluding diary entry (as read by Shelby Foote) which conjures up a highly idealistic and dreamy fantasy where he imagines "reliving the war . . . [with] his fellow soldiers, gray and blue . . . if not here on earth then afterwards in Valhalla . . . [when] the slain and wounded will arise and all will be together under the two flags, all sound and well, and there will be talking and laughter and cheers. And all will say, 'Did it not seem real? Was it not as in the old days?'" Mythic reverie, in the end, then, mixes with the stark historical realities of the series as presented repeatedly over the previous eight episodes in numerous scenes of narrative description which explicitly illustrate the war's brutality through the skillful rephotographing of period images combined with the battle summaries as described offscreen by David McCullough. Even such grandiloquent responses to the war as this final wish fulfillment, moreover, breathe life into history by providing audiences with a far more intimate glimpse into the hopes, longing, and self-questioning of the vanquished as well as the ensuing coping strategies they evidently used to survive and eventually erect the enduring mythological romance of the "lost cause."

The Civil War was subsequently rebroadcast in January and June 1991 and has since been rerun on public television many times. Burns was already back in Walpole, New Hampshire, in early January 1991 editing his soon-to-be eighth PBS special, *Empire of the Air: The Men Who Made Radio* with the other members of his creative team. "I work seven days a week," he admitted then, "almost 15 hours a day. I suffer from the Red Shoes ballet syndrome—you can't stop."[85] The active production phase of *Baseball,* an even more ambitious multipart series, was soon about to begin preliminary shooting during the major league season later that summer. As with *The Civil War* before it, producing a history of the national pastime would explore a defining aspect of coming of age as a boy in postwar American culture during the 1950s and 1960s. Burns likewise contended that "the Civil War literally compelled me to do th[at] film," enabling him to establish "a dialogue with the past." As Barbara Fields reminds us in the final episode of the series, "the Civil War is in the present as well as the past." In this one sense, at least, all history is contemporary. We can never escape our own time or set of ideological predispositions; and within this context, no one had ever before drawn more Americans to history through the power and reach of prime-time television than Ken Burns.[86]

The Civil War, as a result, emerged as the new standard of quality and success by which all succeeding historical documentaries on television would now be judged (including Burns's own later efforts). By 1991, the extraordinary public response to the series had fundamentally changed Burns's professional profile and prospects. His home base in Walpole, New Hampshire, was rapidly transforming from what was essentially a small family business to what would soon become a much more expanded and mature independent production facility involving ten full-time employees, and escalating to as many as thirty or more staff members when editing began on *Baseball* later in 1992. *The Civil War* had culminated a five-year commitment for Burns and his creative associates, including his brother Ric as coproducer, Geoff Ward as head writer, Paul Barnes as supervising editor, Buddy Squires and Allen Moore (along with Burns) as cinematographers, and David McCullough as narrator and story consultant, along with a handful of others.

All of these various individuals had worked with Ken Burns on previous documentary projects and Squires was even one of the founding members of Florentine Films back in 1975 along with another fellow graduate from Hampshire College, Roger Sherman. Fifteen years later, "the dream came true," according to Burns.[87] How they all got to this point now seemed to them as improbable as it appeared to many seasoned professionals inside the television industry. In retrospect, Burns and his fledgling independent company started along the path that eventually led to *The Civil War* after he finished reading David McCullough's *The Great Bridge* on a cold wintry day in January 1977 and immediately decided that he would find a way to produce *Brooklyn Bridge*.

CHAPTER TWO

Life Lessons:
Learning the Basics
on *Brooklyn Bridge* (1982)

EARLY PERSONAL INFLUENCES

I had a case of pneumonia and a friend of mine in January of 1977 gave me
the paperback version of McCullough's history of *The Great Bridge,* and I
suddenly was so inspired. Here was a man who brought history to life. He is
our greatest narrative historian, I think. He let the past speak for itself,
something I had been experimenting with in my own crude college films.

—Ken Burns, 1993[1]

Ken Burns took a giant leap of faith when he fell in love with the idea of adapting
The Great Bridge into an hour-long nonfiction film. Historical documentaries
were neither all that popular with audiences nor the style of choice for most
filmmakers during much of the 1970s. The documentary world was still engaged
in the waning days of direct cinema and cinema verité as the preferred approaches
for most nonfiction veterans. Creating a history of the Brooklyn Bridge on film,
no matter how beautiful and culturally significant the structure remained 105
years after its completion, hardly seemed like the kind of breakout project that
would ultimately launch a newly established independent production company.
No wonder, then, that Burns was greeted with incredulous looks when he left
his sickbed and came into the living room of the apartment he was sharing with

friends and excitedly announced his plans for *Brooklyn Bridge*. Buddy Squires remembers, in hindsight, "we told him to go back to sleep."[2] Apparently the other members of Florentine Films would need some convincing before the idea finally gained a working consensus among them.

Burns had grown up with a special appreciation for the past from his earliest days. His father, Robert Kyle Burns, was an anthropologist who was finishing his graduate studies at Columbia University when Kenneth Lauren Burns was born in Brooklyn, New York, on July 29, 1953. His mother, Lyla Smith (Tupper) Burns, was a trained biologist who now turned her full-time attention to her young family after the birth of her first son. When Ken was only four months old, in fact, they all traveled to Saint Véran in the French Alps so Robert could conduct research and observe firsthand the profound changes occurring within this remote village caught between its medieval ancestry and the inexorable arrival of modern life. A few years later, he even published a short article about their ten-month stay in Saint Véran in *National Geographic*, accompanied by many of his own black-and-white photographs. Upon first arriving, for instance, he recounted how "the warmth of the inn at Saint Véran, the friendly voices of the proprietor and his wife worked magic on our moods. From the kitchen came the aroma of rabbit ragout. Kenny would soon be asleep in a downy bed. I sat back to contemplate the days to come, when we would meet the folk of Saint Véran. The modern age was behind us."[3]

Ken had actually "intended first an academic career in anthropology and then sort of switched and decided [he] wanted to be a filmmaker."[4] According to his brother, Ric, their father was, moreover, "an obsessive photographer."[5] Ric (a family nickname shortened from Eric) was born 18 months after Ken, following the family's return to the United States, while they temporarily lived in Baltimore where a number of relatives resided. Robert accepted his first full-time teaching appointment at the University of Delaware in 1955. Soon thereafter, Lyla was diagnosed with breast cancer which eventually proved metastatic. Ken recalls, "her cancer was the great forming force in my life, permanently influencing all that I would become."[6] Robert later obtained a faculty position in cultural anthropology at the University of Michigan in 1963. A year later, Lyla died. Ken was 11 and Ric was 10. Robert remembers, they "had a very hard time with that." Ken adds, "I've been told by my [former] father-in-law, a psychologist, that my films are a way of, in effect, waking the dead, that maybe I'm trying to keep her alive through the films."[7]

Ann Arbor was a culturally rich and diverse university community during the 1960s. It was especially congenial to a young boy with a burgeoning interest in movies, having an old-style downtown revival theater, which played a wide assortment of classics and contemporary imports, as well as hosting an annual film festival of national reputation and importance. Ken also remembers "when I was growing up my father used to let me stay up late and look at films. I mean he had a strict curfew but for some reason I could stay up to two or three o'clock if Howard Hawks's *Rio Bravo* was on."[8] Like many of the so-called film generation, Burns became acquainted with the work of Hollywood's greatest filmmakers on late-night television; he remembers John Ford, in particular, and the impact of watching *Young Mr. Lincoln* (1939), *My Darling Clementine* (1946), and *Fort Apache* (1948) for the first time.

Ford animated these specific feature films with a deep commitment to the frontier spirit of rugged individualism, nation building, and traditional Christian virtues. Throughout all of his work, Ford established himself as a visual poet of the first order; he was also a sentimentalist and a populist, stressing a sense of nostalgia and a firm commitment to the ways of the past. Many of these conventional elements are still very much a part of Ken Burns's documentaries as well.

> I had always wanted to be a Hollywood director. I looked up to Hitchcock and Hawks and Ford as sort of beacons of how I'd want to do it. But I think as I look back now, in retrospect, I realize how influential Ford was in that if you look at my whole body of work, it's a kind of documentary version of Ford that is a real love for biography, a real love for American mythology, a real love for the music of the period, a real love for ordinary characters who coexist not just on the fringes of our main characters' actions, but who are actually central to the drama and remind us that the best history is not just from the top down, but from the bottom up.[9]

Ken's father bought him a super-8 camera when Ken was a teenager and he began taking motion pictures and assembling his first rudimentary films, further fueling his dreams of becoming a director. Ken also decided to forego the University of Michigan, which he could attend tuition-free as the son of a tenured faculty member, and instead convinced his father to let him apply to Hampshire College in Massachusetts, a small progressive liberal arts institution which had just opened its doors in 1969. The riskiness of this decision was compounded by the fact that Ken, who had graduated from high school a year early in 1970, also needed to take time off to assistant manage Discount Records in Ann Arbor in order to save the additional money needed to enter Hampshire the following fall. He resolutely followed through with this plan, and his choice of college proved fortuitous for both Burns's long-term goal of becoming a filmmaker as well as his eventual growth as a person and a creative artist. The main reason, he insists, is being "fortunate enough to have a true mentor in college, and that was Jerome Liebling."[10]

Liebling was wholly unlike anyone Burns had ever had as a teacher before. He was nearly the antithesis of the Hollywood director prototype that the 18-year-old student from Ann Arbor was now aspiring to be. Liebling, in contrast, emerged out of the tradition that Walker Evans described as the "documentary style," even joining New York's Photo League in the late 1940s where his principal influences included Paul Strand and Walter Rosenblum, among other highly accomplished and socially committed still photographers. He also learned how to make motion pictures from Arthur Knight, Raymond Spottiswoode, and Leo Hurwitz. As Carroll Hartwell, curator of photography at the Minneapolis Institute of Arts, summarizes, "the aesthetic and moral values of former mentors continue to echo, however subtly, in Liebling's sensibilities . . . the League [especially] was his ethical and artistic crucible, charged with the intense energy and talents of those extraordinary individuals."[11]

Liebling's work exhibits a tough awareness of the teeming dislocations and pervasive artifice in contemporary American life. His subjects range from the stockyards of St. Paul to the streets of the South Bronx, populated by ordinary people, politicians on the stump, and even the haunting remains of unadorned cadavers. One distinctive aspect of Liebling's technique is its almost cinematic quality, revealing a sense of drama and movement within the frame, often incorporating the grammar of motion pictures with his extreme close-ups and striking angles.[12] Ken Burns explains that

> interestingly both of these people, Jerry and Elaine [Mayes, another of his teachers at Hampshire College], are still photographers primarily; their work in film has been tangential, but they made excellent film teachers, and guided me . . . I think the amazing thing for us was that film and photography were being taught together, which seems sort of obvious, but I don't know of any other instance where they are. And so there were essentially men and women who had a healthy respect for the image influencing us documentary filmmakers, in fact, persuading us sometimes Hollywood-headed filmmakers that the documentary world could be as dramatic and as revealing as anything that Hollywood can turn out. And I really believe that's true, and I combine that with a latent interest in history to sort of set me on my way. But I think more than anything, Jerry brought a learned wisdom about how you photograph: what you saw, who was seeing, what kind of social responsibility you had at that moment, what you were giving back to the scene that you were, in Susan

Sontag's words, appropriating. All of these things combined with a kind of gentleness of spirit and a good humor that combined to, I think, arm me in the best possible way for the real world and give me a friendship with a man that I still count among my closest.[13]

Ken Burns returned to Hampshire to deliver a testimonial at a ceremony honoring Jerome Liebling's retirement from the college on March 31, 1990. That evening he recounted how "Jerry . . . taught us to respect the power of the single image to communicate, whether in film or photography," and "demanded that we adopt these principles . . . of . . . respect and concern for the subject, strong composition and formal appreciation, and dynamism within the image . . . the honorable practices of still photography."[14] These remarks suggest how Burns fully incorporated the more formally rigorous, socially conscious, and noncommercial standards of the documentary style during his college education at Hampshire, merging these norms with his initial Hollywood aspirations. Ken Burns's future work, as a result, eventually integrated these two very different and largely separate traditions, demonstrating a distinctly personal mixture of artfully serious and popular aesthetics. His ensuing historical documentaries, at their best, and through the apt though unexpected forum of prime-time television ultimately fuse together nonfictional stylistics, historical understanding, and mainstream storytelling techniques reminiscent of epic big-screen moviemaking for tens of millions of Americans.

During his years at Hampshire, then, Burns and his peers were being challenged to consider alternative approaches to motion pictures and photography, while also being grounded in the basic technical and entrepreneurial skills of making a living in media production once they graduated. Burns was part of a close-knit and dedicated group of students who loved creating and discussing films and who "constantly pushed each other," according to Liebling.[15] "Ken was well-steeped in film history," remembers Morgan Wesson, who was a year ahead of Burns at Hampshire and later worked with him on *Empire of the Air;* "he could quote you chapter and verse about the French New Wave, various documentary movements, whatever styles had an impact." In characterizing Jerome Liebling's influence on them all, Wesson adds, "Though he might respect the craftsmanship of Hollywood films, he wasn't about to give an inch. He was trying to convert us all to a private vision, to get us thinking on our own track. Ken took the lead from Jerry and started making documentaries."[16]

Burns worked part-time at the college bookstore during his four years at Hampshire (1971-1975) to help finance his education and subsidize several student film projects while earning a degree in film studies and design. Liebling, moreover, encouraged his students to establish their own nonprofit company called Hampshire Films so that they would be in a position to hire themselves out at no wages to area companies and public institutions who would utilize their maturing talents while underwriting the entire cost of these commissioned productions. Clients were thus able to secure competently made informational films which they could not afford in any other way; and the students, more importantly, obtained a significant amount of much needed real-world experience. As Burns recalls, "it made it possible to leave Hampshire and have the confidence to start my own company and not spend years mired in someone else's vision of things."[17]

Burns's first film as a sophomore, *Bondsville* (1972), was a typical 22-minute student documentary about the decaying fortunes of an old New England mill town, much like many such communities that followed the decline of nearby Springfield, the largest city in western Massachusetts. As a junior, he received a Director's Citation from the Sinking Creek Film Festival for *Two Families* (1973), a study of child abuse funded by the area's Children's Protective Services. He, additionally, experimented with avant-garde and fictional strategies in several of his early short works, one of which, entitled *Yan's 400 Foot Movie* (1974), won notice at the National Student Academy Awards and the New England Film Festival. Burns's budding self-confidence and initiative also emerged in his role as producer of *Transfusion* (1974), sponsored by the Springfield Chapter of the American Red Cross, when the then-ponytailed student filmmaker convinced the touring members of the rock band Grand Funk Railroad to appear in his 10-minute appeal for blood donations aimed at older teenagers and young adults.

Ken Burns's culminating experience at Hampshire College, however, was his thesis film, *Working in Rural New England* (1976), a 27-minute educational documentary which is still used in the visitors' center of Old Sturbridge Village, a living history museum located ten miles southwest of Worcester, Massachusetts. Burns produced and directed this examination and celebration of nineteenth-century farm life, employing many costumed interpreters in re-creations which pictured them husking and kerneling corn, shearing sheep, cleaning, preparing, and spinning wool, milking cows, making cheese, molding pottery, and forging horseshoes and other iron necessities with a hammer and anvil beside a blazing furnace. *Working in Rural New England* was an American Society of Cinematographers finalist for Best Photography in a College Film in 1975 and won a National Historic Preservation Trust film award the next year. The look of Burns's first historical documentary is also strikingly reminiscent of Jerome Lieblings's warmly accented photographs of agrarian life (in contrast to his starker and more unflinching view of urban subjects). As Liebling writes, "having spent so much time among academics, I have a romantic vision of farming, and a deep reverence for those who do this work . . . farming transforms the landscape with colors and shapes, beautiful and dramatic: it is, I think, permissible to romanticize this vision."[18]

In *Working in Rural New England,* Burns, most notably, interjects a series of voice-overs which deliver the actual words of farm women from the era, although he does not yet develop any of these diarists as specific characters within his short historical narrative. He also employs rephotographing for the first time in any of his films, as his camera slowly pans across a large sprawling mural of a New England village from the 1830s, visually spotlighting a number of important details, including an assortment of people milling about period houses, craft shops, farm buildings, and rolling fields and rivers. The experience of producing *Working in Rural New England* was literally a revelation for Burns.

> I began to realize that I had a completely latent and untutored interest in American history. I was passionately concerned with the stories that had animated our national experiment in ways that went beyond academic or scholarly fascination and they [filmmaking and history] just came together. I found myself in a senior thesis at Hampshire College producing a film for Old Sturbridge Village in which I used first-person diary quotes, and I had exquisite live photography at every time of day. And when I got out of college I knew what I wanted to do.[19]

Ken Burns's four years at Hampshire, in hindsight, had greatly expanded his filmmaking horizons and profoundly affected the kind of documentary stylist he was becoming. He, most importantly, had come under the wise and caring tutelage of Jerome Liebling and other working artists, such as Elaine Mayes and her husband Bill Arnold, who moved easily between still photography and film. They, in turn, imparted to all of their students a new appreciation for the single image rather than having them always think in terms of motion pictures as the only way to tell a story. Burns also formed a number of close and abiding relationships with several other highly committed and talented Hampshire students, especially Buddy Squires, Roger Sherman, and Amy Stechler (whom he later married). These friends and long-time associates would soon join forces as company partners once they had all graduated and started along their professional paths.

Burns, in the end, greatly enhanced his technical skills and formal repertoire by attending a small college where he had access to recently purchased equipment even as a freshman. At Hampshire, students were encouraged from the beginning to assist one another on various production projects, thus gaining increased experience in a challenging and supportive environment. Burns also acknowledges, "my parents, of course, were instrumental in every way,"[20] but Jerome Liebling "t[aught] me to be human."[21]

LOOKING TO THE PAST FOR INSPIRATION

I would consider *City of Gold* an influential film, and Perry Miller Adato's
When This You See, Remember Me, a biography of Gertrude Stein, but I think
in both cases I've evolved something on my own.

—*Ken Burns, 1993*[22]

Ken Burns's documentary stylistics appeared refreshingly new to many viewers when *Brooklyn Bridge* first premiered on public television in 1982. His approach to nonfiction filming and formal aesthetics is, nonetheless, based mainly on strategies and techniques that first gained currency in the 1940s and 1950s, and in the case of his attitude and approach to the topics he chooses, the precedents are even older. In retrospect, the nonfiction filmmaker he most resembles in outlook and sensibility is not an innovator from the early days of television who rephotographed a wide variety of still images but, interestingly, Robert Flaherty, the acknowledged father of the form. The resemblance is especially strong when one considers Burns's "left-hand, right-hand process" of simultaneously shooting and writing his scripts, "leav[ing himself] open to discovery as long as possible."[23]

Burns's historical documentaries, of course, do not resemble the much less formal and looser structures of Flaherty's *Nanook of the North* (1922) or *Man of Aran* (1934) or *The Louisiana Story* (1948) in any literal fashion. Still, his use of poetic stylistics, his empathetic (and sometimes romantic) approach to his subjects, and the anthropological spirit of search and exploration that underlies all of his projects suggest Flaherty's own ritual of discovery, or the striving for an openness and full absorption with the topic, more than the stated intentions of any other documentarian, past or present. Flaherty, in practice, would shoot tremendous amounts of footage and then patiently watch his projected film over and over again, searching for new details with each subsequent viewing, eventually shaping his deeply personal and strongly humanistic narratives out of the vast quantities of raw material that he had gathered for the screen. Burns also pays a great deal of attention to the seemingly fleeting and intimate details in his historical subjects, details most people may not see or initially appreciate, often encouraging millions of people to look at old and forgotten archival pictures, for example, in an entirely fresh and different way.

Ken Burns's work, as a result, is probably best known for its still-in-motion cinematography which integrates a variety of single images—daguerreotypes, prints, paintings, and especially photographs—as the primary source material in his historical documentaries. This visual strategy actually dates back to 1940, when it was first employed with distinction in *Michelangelo* (retitled *The Titan* in the United States) by Swiss director Curt Oertel, who lovingly filmed many of Michelangelo's more famous paintings, such as the ceiling of the Sistine Chapel, along with his many landmark creations in architecture and sculpture. Despite Oertel's work, still-in-motion filming continued to be underutilized through the mid-1950s, due to the mistaken notion that this kind of imagery was inherently uncinematic and thus irrelevant to the needs and consideration of most documentarists.

Television, in fact, became a proving ground for the eventual realization of the still-in-motion technique. Sponsorship for compilation documentaries increased after NBC's critical and popular success *Victory at Sea* (1952-1953), a 26-part series on the U.S. Navy during World War II, which was produced by Henry "Pete" Salomon and featured nearly 13 hours of archival film. Salomon and his Project 20 unit specialized in tackling twentieth-century historical topics such as *The Great War* (NBC, 1956) and *The Jazz Age* (NBC, 1957) through the use of stock footage, only resorting to photographs when motion pictures were completely unavailable.

Everything changed with the release of Colin Low and Wolf Koenig's *City of Gold* (CBC, 1957), a Canadian Film Board production chronicling the rise and fall of Dawson City in northwest Canada during the legendary Klondike Gold Rush of the 1890s. Besides short filmed scenes at the beginning and end, the 23-minute *City of Gold* is composed chiefly of 200 black-and-white stills, rephotographed from an animation stand to reveal one moving detail after another, and sequenced dramatically to accompany period music and a spoken narration by writer and one-time Dawson City resident Pierre Berton. *City of Gold* was immediately recognized as a stylistic breakthrough, winning many honors including an Academy Award and spurring other documentarists to incorporate still-in-motion cinematography into their filmmaking repertoire.

The producer-director who best fulfilled the promise of still-in-motion dramatization prior to Ken Burns was Donald Hyatt, Henry Salomon's assistant and eventual successor at Project 20. When Salomon died in 1957, Hyatt continued making compilation documentaries, shifting Project 20's attention more in the direction of nineteenth-century themes such as *Meet Mr. Lincoln* (NBC, 1959) and *Mark Twain's America* (NBC, 1960), each of which contained several hundred photographs, prints, and paintings in the recounting of these historical subjects.

Hyatt's crowning achievement was *The Real West* (NBC, 1961), which was a vast popular success garnering a 42 share in its initial broadcast and eventually winning fifteen national and international awards, including the coveted Premio d'Italia for best documentary in 1961. *The Real West* displayed the many talents of the Project 20 unit. Don Hyatt, along with Daniel Jones and his research staff, examined more than 10,000 photographs, while Philip Reisman, Jr. wrote a lively and entertaining account, peppering a narration filled with stereotyped conflicts (e.g., gunfights, cowboys versus Indians) and predictable personalities (e.g., Billy the Kid, George Armstrong Custer) with an occasional reminiscence from "those who wested," according to the narration. Hyatt's piece is more romance than history, although his close fusion of sound and picture effectively weaves the spoken word, delivered by Gary Cooper, and a period score, composed by Robert Russell Bennett, with over 400 black-and-white photographs and paintings, thus telling a 52-minute story largely through stills-in-motion.

The stylistic contributions of Oertel, Low, Koenig, and Hyatt were often imitated—especially after Hyatt's last major effort, *End of the Trail* (1965) on the American Indian—usually in minor scenes where photographs-in-motion became a well-worn cliche for evoking a bygone era in literally dozens of documentaries as well as fiction films, such as *Butch Cassidy and the Sundance Kid* (1969) and *Days of Heaven* (1978). Eventually Ken Burns eclipsed his documentary predecessors, however, by the formal complexity of his still-in-motion filming, the rigor of his historical research, and the increasingly ambitious sweep of his productions. He, moreover, incorporated an even greater number of technical elements than had ever been attempted or even envisioned before him. Burns recalls,

> I remember I saw a film at Hampshire College about Gertrude Stein, made by a woman—Perry Miller Adato—who worked at channel 13 (public television station, Thirteen/WNET, in New York). She made a film called *When This You See, Remember Me* about the life of Gertrude Stein. It had more theatrical first-person presentations, but it was out of that that I realized that you could take from the past not only its visual evidence—an old photograph, newsreel, a painting—but you could also go back to the printed page and help jump start the past through the eloquence of words. And that was the genesis, that was the seeds for me.[24]

After completing *Working in Rural New England,* which included his first rudimentary attempts at still-in-motion cinematography and dramatic off-screen readings from period diaries, Ken Burns next took the even more ambitious step of creating his own independent production company with two of his good friends and classmates, Buddy Squires and Roger Sherman. They used Hampshire Films as a model after graduating from college in 1975, choosing a "self-initiated route," according to Burns, "a kind of mode of inquiry that [we] brought into the real world, so instead of apprenticing . . . [we] started [our] own company." At the time, they were working as crew members on a cinema verité documentary entitled *In Manhattan* (1975), which their Hampshire College professor Elaine Mayes was producing and directing. On a lark, Burns christened the name of their new filmmaking collective Florentine Films, after Florence, the section of Northampton, Massachusetts, where Mayes lived with her husband, Bill Arnold, an experimental photographer. "We wink and like to say," he joked later, "that it's a renaissance in filmmaking."[25]

Armed with an abundance of youthful energy and idealism, Burns, Squires, and Sherman struggled for several years doing freelance assignments whenever possible, especially as cameramen on a string of visiting nonfiction productions for the BBC and RAI (Italian Television) and on various American documentaries, industrial films, commercials, and a few low-budget theatrical pictures. Burns was hired, serendipitously, by Jean Mudge, who began producing in 1976 a historical documentary on the life of American poet Emily Dickinson. Since 1965, Mudge had served as curator of the Emily Dickinson House in nearby Amherst, Massachusetts, and she was now initiating a career change by founding the self-titled Jean Mudge Productions. Her first major project was *A Certain Slant of Light* (1977), a filmed biography of Dickinson named after one of the poet's more beautiful and haunting verses. Mudge enlisted the participation of Tony award-winning actress Julie Harris, who was flush from her stage success as *The Belle of Amherst,* a one-woman performance which integrated excerpts from poems, letters, and diaries in creating a dramatic portrait of Emily Dickinson. Julie Harris similarly agreed to host *A Certain Slant of Light,* as Burns worked behind the scenes in a number of technical capacities on the production. As luck would have it for Burns, he and Julie Harris also became fast friends. "Ken Burns is very gifted in a human sort of way," explains the actress, "it's all magic with him. He's like an extraordinary actor in that he remembers quotes verbatim. He has such a grasp on the background of his subject and is so interested in every facet."[26]

Burns, likewise, recounts,

I worked as a crew member on a documentary in the seventies in which [Julie Harris] was the narrator-host, and we hit it off and I told her I was going to make a film about the Brooklyn Bridge. She said, let me know if I can help, and I called her back and she came into the studio and read the voice of Emily Roebling, the wife of the man who built the Brooklyn Bridge, and she was terrific. And so I used her name and I called up Kurt Vonnegut and I said, "well we've got Julie Harris," and he said, "well if you've got Julie Harris," and he agreed, and Arthur Miller agreed. We use the same people over and over again, and they've become like our crew and—like the other people who work with us—part of the family.[27]

Full preparation started on the Brooklyn Bridge Film Project in the summer of 1977. Burns, Squires, and Sherman were still the only full-time members of the Florentine Films family at that point, and their biggest challenge was securing the necessary financial and professional backing to produce the kinds of films they aspired to create. Attracting support for *Brooklyn Bridge,* Burns declares, was "the most difficult obstacle of my life."[28] He approached literally hundreds of potential public and private sector underwriters; the few who would listen to him tried to persuade the young producer-director to reconsider his hour-long plans and, instead, film a more manageable ten-minute history of the bridge.

David McCullough, too, first remembers being approached by Ken Burns at an academic conference where the 24-year-old unknown and largely untried filmmaker expressed an intense interest in creating a documentary on *The Great Bridge:* "I didn't want to have anything to do with him, and if anybody was going to make a film based on my book, I wanted it to be somebody who had more experience and more standing."[29] To this day, however, Burns contends that "perseverance is the single greatest element" in his success. He wrote McCullough letters, and later he and his associates contacted him by phone. "The people who work for me, we work hard, we don't give up."[30] McCullough finally acquiesced. As Jerome Liebling explains, Ken "had direction and tenacity, but he looked about 12 years old. People scorned his youth—who is this kid?— but as soon as he began to make films, that attitude changed."[31] "We tested our mettle in every aspect of filmmaking," Burns concludes about *Brooklyn Bridge,* "we developed all the skills used in subsequent projects."[32]

DO I HAVE A DEAL FOR YOU

When I was working on the *Brooklyn Bridge* film I was 23, not quite 24, and I looked about 12. And thousands of people would slam the door and say, "Ha!

This child is trying to sell me the Brooklyn Bridge!" And it was humiliating and for 10 years I kept on my desk, and I still have them in a bookcase next door, these two gigantic binders filled with rejections for the *Brooklyn Bridge* film. The first money I got was from the New York [State] Council for the Humanities, which was a challenge grant [requiring matching funds], and I literally went around collecting $1,000 gifts from people and New York institutions and corporations and put it together. It's always been difficult.

—*Ken Burns, 1996*[33]

Brooklyn Bridge took Ken Burns almost five years to produce—roughly from the summer of 1977 through the spring of 1981—or just a few months less than the time it took to make the much longer, more expensive, and increasingly ambitious miniseries *The Civil War*. Burns faced a chronic stream of rejections from potential investors in attempting to finance the Brooklyn Bridge Film Project (as it was formally presented to contributors), interspersed with occasional flashes of encouragement from people interested in his subject and with just enough financial support to keep the whole venture precariously afloat, often at moments when it appeared that the entire production would have to shut down for lack of support. Tom Lewis, an eventual research collaborator and consultant on the film, recalls that Burns "would go around to New York Telephone, Con Ed [Consolidated Edison, energy supplier], these places, and they'd say no, and he'd be back the next week, and they'd say no and he'd be back again. Then he goes and makes a brilliant film, and it was because of the extraordinary energy that he had. It was like nothing I had ever seen."[34]

Burns received his first big boost in producing *Brooklyn Bridge* when he attended a Brooklyn Bridge Symposium at Skidmore College in upstate New York during the first week of October 1977. It was there that he met Tom Lewis, an English professor who had spent months organizing this two-day affair, an offshoot of his long-time teaching and research interest in Hart Crane, whose most ambitious book of poems was *The Bridge* (1930).[35] The conference turned out to be a significant gathering of leading Brooklyn Bridge experts, including social philosopher and educator Lewis Mumford, American studies scholar Alan Trachtenberg,[36] and David McCullough, who had published his best-selling cultural history, *The Great Bridge*, just five years before.[37] Three documentary teams also attended, each planning a filmic adaptation of McCullough's book. Burns wisely arrived a week early with Buddy Squires and Roger Sherman in order to meet Tom Lewis and make advance preparations to shoot selected photographs, prints, and paintings collected for the symposium and displayed prominently in a large meeting room at Skidmore. Lewis recalls how surprised he was when he initially saw the boyish-looking filmmaker for the first time, but his skepticism soon subsided as he watched Burns "walking around talking to Buddy about how he's going to run the camera over these pictures and I thought—this guy is seeing this stuff the way I do."[38]

"Ken actually was a great help in the conference," Lewis recounts, "and it was a very good conference for him because he made the connections he needed to make in order to get into the archives of New York, to get started with McCullough at least, and to get to meet Lewis Mumford."[39] Mumford, in particular, took an immediate liking to Burns. Although inexperienced, the young producer-director radiated personal charm and a passionate enthusiasm for the film he wanted to make. Mumford, in turn, became a kind of mentor for Burns, advising him on how to proceed, what to think about, and who to contact next. He even gave the filmmaker his "very first interview—the very first roll one of my first film was that moment on October 5, 1977 at the Surry House at Skidmore College and he responded so generously." "As we got smarter about what we wanted to do," Burns adds, "he invited us back to his house and we interviewed him two more times in 1979 and 1980 . . . Lewis Mumford was really there at the very beginning and I think set in motion a career that has followed a lot of his words."[40]

The nearly five-year production process of *Brooklyn Bridge* was, indeed, an arduous endurance test for Burns and his colleagues at Florentine Films. Looking at it another way, though, all the time-consuming challenges they faced provided them with a generous and elongated learning curve on which they had the time to develop technically by mastering the mechanics of their craft through trial and error, and to mature personally and creatively by listening and learning from two highly talented and well-seasoned role models. Mumford, first of all, passed on to Burns a richer appreciation "that you could find, in stone and steel, a history of a great city." He also guided the inexperienced producer-director through some of the thornier issues of proposal writing and willingly composed sterling references for him. "He wrote a letter to Joe Duffey, who had been a student of his," Burns discloses, "who was then chairman of the National Endowment for the Humanities, urging him to take a chance on this completely green and unknown filmmaker; and the National Endowment for the Humanities broke ranks and gave me a grant that allowed me to finish this project, thanks to Lewis Mumford."[41]

David McCullough, additionally, served as a script consultant on *Brooklyn Bridge*, essentially coaching Burns in the art of historical storytelling. McCullough is part of a long popular tradition which includes thoughtful, readable narrative historians such as Shelby Foote, Paul Horgan, Barbara

Tuchman, and Wallace Stegner, all conducting their research and writing mainly outside of the institutional realm of academic history. He portrays richly drawn and compelling historical characters, and emphasizes the inherent drama of those seminal American events he chooses to write about from the country's past. McCullough, most importantly for Burns, employs first-person commentaries "when I can," integrating "the very words used" from letters, journals, diaries, and newspapers. "I want you to feel what happened. I want you to feel the atmosphere of that time," he explains. "The importance of narrative writing is to put you there, not tell you what to think."[42] Ken Burns, of course, was already experimenting with similar techniques on film; and given his genuine love for American history, the young producer-director couldn't have asked for a better script advisor at a more opportune time.

> Here was a subject [the Brooklyn Bridge] that seemed to be a part of the hidden history of America, the history we were never taught, as we focused on wars and presidents and Indians fighting and lawlessness; here was something going on that seemed to speak as much about who we are as anything, and it was urban and it was eastern and corrupt and dangerous. It was about the arts and the sciences that would be more influential in the twentieth century than a lot of the mythology of the nineteenth century, and I sort of went at it wholeheartedly. And more than that, as I got to know McCullough, he became very helpful in refining a story and clarifying how you tell a story. And I think if you combine this great visual and sort of honorable teaching of Jerry Liebling with McCullough's sense of narrative, that's a pretty potent combination.[43]

McCullough, overall, provided Burns with a prototype for writing and producing history, as the young filmmaker fine-tuned his own storytelling talents during the late 1970s and early 1980s. The 57-minute narrative of *Brooklyn Bridge* is essentially divided into two parts. The first and longest section, lasting 37 minutes, recounts the epic saga of the construction of what was at the time the largest and most stunningly beautiful steel-wire suspension bridge in the world. McCullough's first popular history, *The Johnstown Flood* (1968), had told the story of the 1889 Pennsylvania disaster where over 2,000 people were killed in less than 10 minutes.[44] "After the astounding human short-sightedness of that terrible flood," McCullough notes in explaining why he turned his attention to the Brooklyn Bridge, "I wanted to write about a symbol of affirmation of people doing things right."[45] Neither his book nor the filmed adaptation ever grows excessively maudlin or melodramatic in the telling, however, despite the heroic tone and mythic struggle intrinsic to both. Each provides the necessary balance between nineteenth-century American virtues of personal initiative, ingenuity, and succeeding against great odds with similarly evident vices of unbridled greed, fraud, and abuse of power. "Boss Tweed had his hands in the building of the Brooklyn Bridge," Burns relates on why he, like McCullough, was intrigued by the story, "we corrupt as much as we construct."[46]

The young filmmaker, consequently, followed the more experienced example of the author, even launching the first 20 seconds of *Brooklyn Bridge* with a voice-over by actor Paul Roebling (a direct descendant of both the designer and the engineer who was in charge of the building operation) reading the same epigraph which started *The Great Bridge:* "*Harper's Weekly,* 1883, 'It so happens that the work which is likely to be our most durable monument, and to carry some knowledge of us to the most remote posterity, is a work of bare utility; not a shrine, not a fortress, not a palace, but a bridge.'" The story, then, unfolds with David McCullough, also serving as narrator, describing the *Brooklyn Bridge* as "a work of art" and "the greatest feat of civil engineering in the world." He relates that it was the "inspiration" of a "Renaissance man," John A. Roebling, who died as a result of an accident just as the building of the bridge was in its earliest stages, and his son, Washington Roebling, who finished the monument 14 years later through his own dogged persistence, despite being bedridden in the process by caisson disease (known today as the bends). Washington's wife, Emily, also figures prominently throughout the historical narrative, courageously serving as her husband's surrogate to the outside world during the 11 long years he is literally too sick to leave their apartment in Columbia Heights overlooking the bridge.

After this 8 minute and 15 second introduction to the scope of the subject, the principal characters, and the general plot line, Ken Burns next organizes the narrative into seven successive chapters (i.e., "Early Bridges," "Caissons," "Towers," "Politics," "Cables," "Roadway," and "Opening Ceremonies"), which is again highly reminiscent of *The Great Bridge* in abridged form. This method of segmentation is now representative of Burns's historical documentaries in general, as is the overriding biographical focus on the Roeblings, a number of their principal antagonists (e.g., William Marcy Tweed, J. Lloyd Haigh, the New York press) and an assortment of other colorful figures (e.g., Mark Twain, Georg Friedrich Hegel, Walt Whitman) who make strategic cameos throughout the first two-thirds of the nonfiction story. Part two (or approximately the last 20 minutes) explores the complex and sometimes contradictory symbolic meaning of the bridge, featuring extended commentaries by Lewis Mumford, David McCullough, playwright Arthur Miller, architectural critic Paul Goldberger, Hart Crane scholar Jack Unteracker, and Henry Jones, a 106-year-old man who actually worked on the construction of the bridge. As Burns explains in the report that he made to the National Endowment for the Humanities on completing *Brooklyn Bridge*,

> the bridge becomes the personal embodiment for those interviewed of a very serious and powerful structure: for some a heroic patriotic event, a symbol of perseverance against all odds, the embodiment of the American myth of self and destiny; for others an almost cosmological symbol of the coming together of one Thing of great beauty and utility symbolic of the highest humanist ideals, the Bridge as transcendental object; still for others the bridge is the archetypal American work of art, rooted in technology, yearning for immortality.[47]

Part two, in this way, also surveys the many serious and humorous instances in which the bridge has inspired artists (e.g., Joseph Stella, John Marin, Georgia O'Keefe), photographers (e.g., Berenice Abbott, Walker Evans, Edward Steichen), Hollywood films (e.g., *It Happened in Brooklyn, The Bowery, Tarzan's New York Adventure*), and even commercial advertisements (e.g., Vaseline, Singer sewing machines, Dr. Brown's natural flavor celery soda). Burns deftly combines differing categories of information and various modes of aesthetic expression: specifically, *Brooklyn Bridge* is a mix of weighty and more light-hearted content, modernist and popular stylistics. The young producer-director was aiming for an approach and "context accessible to most," even with this first major production.[48] His stated intention of "bringing the past back alive on screen" was analogous to David McCullough's aforementioned goal of "put[ting] you there" in his historical writings and fully consistent with the higher ideals of his mentors, such as Lewis Mumford's desire "to see something larger" and Jerome Liebling's insistence on preserving "the observational integrity of the single image."[49]

Liebling's curricular linkage of film and photography at Hampshire College, for example, now emerges in *Brooklyn Bridge* as one of Burns's most unique and identifiable stylistic trademarks: he treats old still images (i.e., photographs, paintings, bridge blueprints) as if they were moving pictures, panning and zooming within the frame, shifting back and forth between long shots, medium shots, and close-ups; while correspondingly handling the many live shots of the bridge as if they were still photographs. Burns's own cinematography in the film, taken at all times of the day and during every season, is characteristically formal and painterly, almost in an academic sense. This emphasis on static composition is particularly effective in evoking the mood and prefilmic visual vocabulary of the nineteenth century, corresponding directly to the historical era and topic he is exploring in *Brooklyn Bridge*. In one extraordinary instance, he recalls

> going out one rainy day, and we weren't really sure why we were out there. It was sort of whitish-gray and terrible. All of a sudden, the sky opened up in the west and we took a series of shots. I still remember it to this day—this was in 1979—May of 1979—it was roll 53. We took a series of shots—six or seven shots in the three or four minutes that this light was there. I have never in my life seen light duplicated this way. It was so stunning, so hard, so beautiful with an orange western light, gray, white, but incredibly defined clouds. What it did to the light on lower Manhattan, what it did to the tugboats coming up, what it did to the bridge was unspeakable, and I can remember that as an epiphany of filmmaking.[50]

By almost any measure, in summary, Ken Burns's now distinctive and well-recognizable style is conservative in the sense that it relies entirely on influences and techniques first introduced decades ago; he, however, arranges these constituent elements in a wholly new and highly complex textual arrangement. Beginning with *Brooklyn Bridge* and becoming increasingly more integrated and better defined thereafter, Burns blends narration with what he calls his "chorus of voices," meaning readings from personal papers, diaries, and letters; interpretive commentaries from onscreen experts, usually scholars, critics, and witnesses; his rephotographing technique which closely examines old photographs, paintings, drawings, and other artifacts with his movie camera; all backed up by sound effects and a music track that features period compositions and folk music. The effect of this collage of techniques is to create the illusion that the viewer is being transported back in time, literally finding an emotional connection with the people and events of America's past.

Burns's style, significantly, also demonstrates his resourcefulness in turning what are often matters of inherent adversity for most underfunded independent filmmakers to his own creative advantage as a producer-director. All of the archival images he employed in *Brooklyn Bridge,* for instance, were part of the public domain and, thus, not subject to the high fees typically charged for permission and use. Burns likewise scoured used record stores and flea markets as a more cost-effective way of finding old sheet music and historic recordings which comprise most of the incidental music on the soundtrack. This practice, which he and his colleagues repeated on all of his films over the next decade, not only saved money for Burns but also exhibited artistic benefits as well. In *Brooklyn Bridge,* for example, eight out of the nine tunes used are traditional in origin (i.e., "Old Rosin the Bow," "Sweet Home," "The Union," "Longshore-man's Blues," "We Shall Be Happy," "20th Century Wonder," "President's March," and "Greenfields of America"). For "20th Century Wonder" he even utilizes a version performed by a high school band that, instead of sounding amateurish, evokes an earlier performance standard, much like the antiquated quality of a sepia photograph, adding to the audience's suspension of disbelief as Burns inventively applies the tools of his trade to bring this nineteenth-century American story back alive on screen.

"Work[ing] without salary for the better part of [1980],"[51] Ken Burns was, in turn, able to convince most of the highly accomplished performers and writers (i.e., Paul Roebling, Arthur Miller, and Kurt Vonnegut) who constituted his chorus of voices to either work for union scale or, in the case of Julie Harris, donate her services altogether.[52] *Brooklyn Bridge,* in the end, cost $180,860.50, which was an immense sum of money for a young filmmaker like Burns to have raised between 1977 and 1980, especially since he had no professional track record of note.[53] The key, once again, to his eventual fundraising successes were the contacts that he forged at the Brooklyn Bridge Symposium at Skidmore College in October 1977. That conference, in fact, was funded largely by the New York State Council for the Humanities. A representative from that agency, who soon afterward assumed the directorship, attended both days, and Burns made a point of getting to know her as best he could. Tom Lewis, too, was won over by the enterprising producer-director's enthusiasm and apparent technical compe-tence, and as a result, he helped pave the way, remembering that eventually

> [Burns] gets this $50,000 from the New York Council for the Humanities which was like a gazillion dollars. I mean, here is this kid who had gotten this money because we schmoozed this woman from the New York Council. I can remember saying at the cocktail party that night, "well you've got to give it to this guy who can make this film," and she said, "but he looks so young, when he brought the application in, I thought he was just a messenger boy." And I said, "I know he looks like a 12-year-old, but believe me, this guy has the aesthetic sensibility. I've talked to these other filmmakers, this is the one to do it."[54]

The New York State Council for the Humanities ended up awarding Burns $50,000 on March 28, 1978, with the customary proviso that he was required to raise a matching $50,000. He next relocated to the New York area to continue the difficult and challenging process of procuring more money for *Brooklyn Bridge* and researching the bridge's history. He, moreover, wanted to be nearer to his one-time college girlfriend and now closest filmmaking associate, Amy Stechler, who was moving back to her family's home on Long Island after graduating from Hampshire in 1977 and was completing a year of advanced film study at MIT with Richard ("Ricky") Leacock, one of the founding pioneers of direct cinema. Burns leased an apartment overlooking his majestic subject in a five-story walkup in Brooklyn Heights and started the prolonged and exacting phase of finding the rest of the financing for *Brooklyn Bridge*. He was, additionally, spending large amounts of time learning more about his topic at the New York Public Library and at the Museum of the City of New York Archives, which was supportive enough of the young filmmaker's efforts that staff members provided him with a small space on their premises to use as a makeshift office.

Ken Burns, in many ways, was largely unaware of what exactly he had gotten himself into. His correspondence at the time suggests that he generally expected to be finished with *Brooklyn Bridge* sometime toward the end of 1979. He had official letterhead printed up announcing "The Brooklyn Bridge Film Project," with an image of the bridge taken from a well-known print adorning the top of the stationary. His professionalism and political acumen were evident even at this early stage in the way he presented himself, his company, and his developing film project to the literally hundreds of municipal, state, and national businesses, public institutions, and major foundations that he was regularly soliciting; the rejection rate, in response, exceeded 25 to 1.[55] Burns, ironically, received notices from both New York Telephone and Consolidated Edison during the fall of 1978 informing him that services were about to be shut off in his Brooklyn apartment for debts in arrears, while he was simultaneously lobbying these very same companies for funds to produce *Brooklyn Bridge*. By the end of the year, Burns and Stechler decided to move together out of New York to Walpole, New Hampshire, a quiet, picturesque village on the Connecticut River with a population of 3,210, where the cost of living allowed them to survive on as little as "$2,500 one year to stay independent."[56] "Amy and I moved here in '78," Burns recollects. They converted a modest one-time garage behind the Cape Cod–style house they shared into an editing facility and office. "We cut *Brooklyn Bridge* here [referring to the small building]. This was built in 1879. That stove used to be our heat. Amy and I would come out here and edit film, day and night."[57]

Shooting sessions of the Brooklyn Bridge and at various libraries and museums in the northeast began in 1978 and continued throughout the next year, while Burns doggedly pursued investors as best he could. He finally cobbled together the necessary combination of matching funds by July 1979, slightly exceeding the amount required by the New York State Council for the

Humanities by raising a total of $52,400.[58] Editing also took more than a year, starting in September 1979 with the creation of a 7-minute "teaser" to help garner even more monies and then extending through the remainder of 1980, honing down approximately 30 hours of footage to a rough cut of just over 1 hour.[59] The final stages of assembly followed, including the recording of the chorus of voices and the narration, along with mixing 19 separate soundtracks into one with the incorporation of the traditional songs and some original music composed by Jesse Carr.[60] As late as March 14, 1980, Burns was still describing *Brooklyn Bridge* as "severely underfunded . . . caus[ing] something of a crisis."[61] The National Endowment for the Humanities soon awarded the filmmaker $25,000 (in reply to a $60,000 request); and then Citibank contributed a comparable sum of money later that summer.

By December 1980, Ken Burns estimated that they were "about $20,000 short of the funds needed to complete the film"[62] when "NET [PBS station WNET in New York] offered me that helping hand that brought me out of frigid waters into the life raft."[63] This willingness to extend financial support combined with a genuine interest in both his subject matter and his stylistic approach had a profound impact on the thinking and future plans of the young filmmaker. Burns had always envisioned "my films to be shown on film in a dark room with many people, but it is impossible to get funding consistently for the kind of long-term, labor-intensive projects that I had in mind."[64] Public television provided him with a viable alternative in a highly sympathetic environment. The programming representatives at Thirteen/WNET not only had the resources to help subsidize and telecast *Brooklyn Bridge,* they were also open-minded to the kinds of historical documentaries that he was now most interested in producing and directing. Ken Burns, as a result, began to refocus his professional energies toward television as the primary venue for releasing and distributing his work rather than movie theaters. "In the case of [his contacts at] WNET," he explains, "they gave the finishing costs that allowed the film to be done and then subsequently were part of the formal mechanism for presenting the next couple of films that I made."[65]

Brooklyn Bridge premiered nationally on public television on May 24, 1982, the ninety-ninth anniversary of the official opening of this historic New York landmark. Five years and four months had passed since Ken Burns had left his sickbed in Amherst, Massachusetts, flush with the sudden realization that he wanted to somehow adapt *The Great Bridge* as his first major, professionally produced and directed effort. Much had understandably changed for him during the long and demanding process of seeing the Brooklyn Bridge Film Project through to conclusion. All the hard work and perseverance resulted in nearly uniform praise from audiences and critics as *Brooklyn Bridge* won a series of industry-wide and scholarly commendations, most notably an Academy Award nomination for Best Documentary in 1982.[66] This film was also his first prime-time PBS special and ushered him into the ambit of public television where he has operated and flourished ever since.

"Back then," Burns mused in 1994, "I was telling people that what I wanted in ten years was to be living in the country, making documentaries and married to Amy."[67] He stayed steadfastly focused to this self-scripted scenario, as Amy Stechler reveals that "Ken can remember what he set out to do and follow that straight line, no matter what."[68] In turn, they were wed on July 10, 1982, settling ever deeper into their adopted hometown of Walpole, New Hampshire, and sharing a period together when their personal and professional interests became largely indistinguishable from one another. Burns similarly characterizes the many aspects of producing and directing film as almost a way of life. He maintains that making documentaries has trained him "how to see, how to create, how to write, how to shoot, how to interview, how to criticize, to raise money, to speak in public, and how to research. It even taught me how to tie a tie. It's a total education."[69]

All in all, after *Brooklyn Bridge* Ken Burns was more secure than ever with the career choice he had essentially made as a teenager in Ann Arbor, subject to the adjustments in style and content that he enacted in response to the guidance and advice from his principal mentors, especially Jerome Liebling, Lewis Mumford, and David McCullough, and the scores of feature filmmakers and documentarians whose work he had watched and learned from for years. He had been greatly influenced and supported too by his closest companions and colleagues from Hampshire College and Florentine Films which "was started in 1975," Burns recounts, "and has undergone many metamorphoses."[70] The original three members of the company—Ken Burns, Buddy Squires, and Roger Sherman—expanded to five in 1978 with the addition of Amy Stechler and Larry Hott, a mutual friend who had recently graduated from Western New England Law School in Springfield, Massachusetts. Hott worked in 1977 as a legal services attorney in Oregon and then returned to New England, partly on Burns's urging, to join the Florentine collective and rechannel his advocacy energies into making films. The first project he choose to produce was *The Old Quabbin Valley* (1981), a public interest documentary on the various factions competing for water rights in central Massachusetts. Hott remembers that "it was a very magic, special time for a lot of reasons. One was we were young. We were not supporting families. We were at the time of life where we were saying this is a chance to experiment, and we decided we could do things like live very simply and cheaply to get films made."[71] When the production processes for *Brooklyn Bridge* and *The Old Quabbin Valley* finally ended, however, the old friends and associates at Florentine Films were basically ready for a change. Each was moving personally and professionally in their own slightly different direction. The next immediate challenge facing them, and their last group effort as a full cooperative, was to restructure their jointly held company to better reflect this newly emerging reality.

CHAPTER THREE

Variations on a Theme: American Originals, Symbols, and Institutions

A LABOR OF LOVE

What brings me to these subjects is in the wonderful and peculiar tension between granite and steel in the Brooklyn Bridge, in the grace of a Shaker chair, and in the symbolism of the Statue of Liberty, even in the corrupt demagoguery of Huey Long. All of these subjects are animated by the question, "who are we?"—that is to say, who are we as a people? What does it mean to be an American? All of these questions are not necessarily answered by these investigations as the question is itself deepened, and that's basically, almost in a nutshell, what I do.

—Ken Burns, 1993[1]

The Shakers: Hands to Work, Hearts to God was very much a family affair from start to finish. Having focused all of his attention for more than four years on completing his first large-scale historical documentary, Ken Burns was now uncertain where to next direct his professional energies during the spring of 1981. In his final status report on *Brooklyn Bridge* to the National Endowment for the Humanities (NEH) in May 1981, he wrote: "There are no immediate plans to follow this project up but its phenomenal success has created an interest in pursuing a series of films on American monumental architecture."[2] Burns was apparently in a period of intense self-examination through the ensuing spring and summer months as evidenced by the tone and substance of his work journal entries at the time. In one example he reflected, "how do I nourish the witness? There is so little observation, so little seeing. There is a moment of awakening . . . [then] despair, sleep . . . The moment of possibility is so quick. The need to go deeper and not rest."[3]

Finally in August of 1981, he and Amy Stechler happened upon an early morning sight that eventually led to the start of their next major film project. While driving through the Berkshire hills, five miles west of Pittsfield, Massachusetts, Burns spotted an exquisitely constructed and preserved round stone barn from 1826 gleaming in the sunlight on the left hand side of the road. "I slammed on the brakes and went in," he recalls. What Burns and Stechler had discovered was Hancock Shaker Village, The City of Peace, a living history museum comprised of 20 restored buildings which operated as one of the most prominent Shaker communities in America from its founding in 1790 until its closing in 1960. The then 28-year-old filmmaker reveals that he was immediately struck "by the question: who are these people?"[4] He and Amy Stechler, as a consequence, began a six-year partnership of exploring the Shakers together, which resulted in a 59-minute historical documentary telecast nationally in 1985, followed three years later by a companion book of text and photographs.[5] During this period, their first child, Sarah Lucile was born during the making of the film, and their second daughter Anna Lilly was born during the making of the book.[6] They also forged personal relationships with two of the Shaker eldresses who appear in their film, Bertha Lindsay and Gertrude Soule, whose living example "inspired" the young couple "with their simplicity . . . grace" and the "relevan[ce of the Shaker lifestyle] in the twentieth century as it was in the eighteenth."[7]

Burns and Stechler's newly found interest had been addressed on film several years before in *The Shakers* (1974), by Virginia-based independent producer-director Tom Davenport in collaboration with Daniel Patterson, who teaches at the University of North Carolina's Curriculum in Folklore. Funded by NEH and shown on public television, Davenport's 30-minute documentary was arguably the definitive motion picture portrait of the Shakers up to that time, featuring mostly interviews with now deceased Shaker women, capturing their thoughts about daily life, religion, and the future prospects of their sect. This movie includes recorded performances of Shaker songs by the late eldresses Marguerite Frost of Canterbury, New Hampshire, and Mildred Barker of Sabbathday Lake, Maine. There are also a few rephotographed stills and drawings briefly illustrating the history of the Shakers, but, by and large,

Davenport's work is an informative compilation of talking and singing heads, which is chiefly characteristic of the era's allegiance to cinema verité.

Ken Burns, in contrast, furthered his own singular growth as a documentary stylist during the production of his and Amy Stechler's version of *The Shakers*. They had been "looking for interesting rural architecture" to photograph that morning in August 1981 when they first observed Hancock Shaker Village. "The road was too well traveled for surprises," recalls Stechler, "[i]nstead it was lined with endless shopping centers, motor inns, and car dealers. Suddenly, we drove into the heart of a perfect, early nineteenth-century village. The buildings were so compelling in their unusual shapes and luminous colors that we pulled off the road and spent the day."[8] Burns is a great believer in "go[ing] into the places where the past happened and to sort of listen to the ghosts and echoes of this inexpressibly wise event and to learn something from it."[9] He took dozens of preliminary photographs that day, along with hundreds of subsequent pictures on return visits to Hancock as well as to Shaker sites in Pleasant Hill, Kentucky, and Canterbury, New Hampshire.

Ken Burns regularly acknowledges his attraction to "the very powerful emotional resonances that seem to emanate from the collision of individuals and events and moments in American history."[10] In the case of *The Shakers,* he began his exploration of the topic by acclimating himself at the outset to the *genius loci* (the character of the place), long before he ever began thinking in terms of filmic symbols and a possible narrative structure. As with many creative people, especially those who are visually oriented, Burns initially gains knowledge "sensually, emotionally, intuitively, and privately" before transforming these insights "into public expression using the language of [his] discipline."[11] This producer-director in particular believes strongly in "the power of the single image to communicate," more so than most filmmakers.[12] He likewise explains that "there is something in the still image that is the building block of what we do. It's a sort of blank slate that is really about the first step in these projects."[13]

Ken Burns also identifies directly and intensely with his subjects. A dozen years after finishing *The Shakers,* while filming *Lewis & Clark: The Journey of the Corps of Discovery,* he remembered back to his earlier film on the religious and communal sect as he experienced a startling revelation about his own approach to filmmaking, his enduring preoccupation with the elemental question, "Who are we as Americans?"

> I'm driving along in my Chevy Suburban through this amazing landscape, and
> I'm going 70 miles an hour. I've got 70 miles to go until the next town, and
> I'm just dwarfed by the land . . . I realized that traditional American religious

experiments—the Shakers, the Mormons—thrived on the frontier. That, paradoxically, the time when the question of physical survival was the loudest was also when the question of the soul's survival was the loudest. At that moment I realized that I had deceived myself: my question was not "who are we?" but "who am I?" It was wonderful, because at that moment the question went out and down the valley, down the bridge, over the Missouri, and it echoed through the canyons and it stole back, very quietly, into the car where we were, and settled once again as "who are we?" For a brief moment, the question "who are we?" had united with every individual American's question of "who am I?"[14]

A deeper though no less tangible dimension of Burns's quarter-century commitment to history, in this way, is derived from his continuing introspection as a creative artist which underlies his more obvious public inquiries as a popular historian. When Burns astutely proclaims that the "country [is] starved for national self-definition today,"[15] he is similarly speaking about himself. The producer-director's self-exploration through his documentaries is also the reason why he often refers to the therapeutic nature of history as one of its most salient properties: "I think there's a sense that the past might be some kind of healing or medicinal force, that knowing where you've been arms you in the best kind of way for proceeding forward into the future."[16] Burns was clearly moving on to an entirely new chapter in his life both personally and professionally when he decided to produce *The Shakers* with Amy Stechler, starting with their impending marriage in the summer of 1982, and the imminent devolution of Florentine Films.

The beginnings of an eventual change in the organizational structure at the Florentine cooperative was actually set in motion in 1978 when Burns and Stechler relocated to Walpole, New Hampshire, some 75 miles north of their other filmmaking partners who were primarily based in and around Northampton, Massachusetts. Friend and former business associate Larry Hott, remembers that

we had this sort of utopian, idealistic view of how a film company could run. It was more of a collective concept. Each person was supposed to develop their own film in conjunction with everybody else . . . Ken's decision to move up to New Hampshire sort of meant that we weren't operating completely as a cohesive group, and I don't know what would have happened if he had stayed around and we all had lived in the same town—maybe we would have worked together a little longer.[17]

While Burns and Stechler were editing *Brooklyn Bridge* in Walpole during much of 1980, two other half-hour projects were well underway by other Florentine members. Hott was busy raising money and finishing his first self-initiated film, *The Old Quabbin Valley,* while Roger Sherman was largely preoccupied with completing *Ski New Hampshire* (1981) for the New Hampshire Office of Vacation Travel. Although each member was supposed to be helping out the others on all the pending projects at the company, Burns, Hott, and Sherman were understandably consumed with seeing their own projects through to completion rather than spending great amounts of time on the other documentaries. Buddy Squires was then far more interested in cinematography than producing, so he worked as a cameraman and assistant on the three Florentine films, while also accepting a series of other outside assignments, such as location shooting in Cambodia on *The Khmer Rouge* (1981) for Télévision Français 1 (French network television). Terry Hopkins, another Hampshire College colleague, and Diane Garey, who later married Hott and became his filmmaking partner, also assisted in a number of camera, crew, and writing capacities at the collective.

Still, the informal nature of Florentine Films made determining precise credits more difficult than expected. *Brooklyn Bridge, The Old Quabbin Valley,* and *Ski New Hampshire,* moreover, marked three distinctly different nonfictional approaches (i.e., historical, public interest, and promotional/commercial), representing the somewhat divergent directions in which Burns, Hott, and Sherman were heading. Most importantly, though, the disparity between the number of years that were needed to finish these projects and the amount of money that was available for everybody to live on signaled an inevitable restructuring of the company. They all agreed on how necessary it was to further professionalize their cooperative in order to ensure its long-term viability and success.

The five principal shareholders—Burns, Squires, Sherman, Stechler, and Hott—decided to trademark the name "Florentine Films," indemnify the entire group against any future lawsuits or damages, and decentralize the operation under the aegis of each separate partner. Roger Sherman, for instance, founded Florentine Films/Sherman Pictures in Northampton (later moving his operation to New York City). Larry Hott and Diane Garey created Florentine Films/Hott Productions in Williamsburg, Massachusetts (a small town five miles west of Northampton). Buddy Squires continued doing freelance cinematography under the rubric Florentine Films, while Burns and Stechler remained in Walpole, producing historical documentaries such as *The Shakers*. "I actually have a company called American Documentaries that sort of is the shadow behind Florentine Films," explained Burns, "we essentially disbanded and allowed each other to use the name."[18]

The five associates clearly recognized the practical value of keeping the Florentine title now that the collective was finally building a reputation with each of their first three films winning awards, even though the notion of branding a film company and its product was not yet a widespread practice among independent producers during the early 1980s. "At the beginning," Hott recounts, "we would have a resume or filmography of 12 films for the three of us . . . [so when] we would go to people for funding, they would look at it and say, 'this is a large company,' and we trusted each other to maintain a high aesthetic standard for our films. People are very careful about what comes out of our individual shops."[19] After the agreement was formalized in 1982, Burns notes that there were "four incarnations of Florentine Films around New England."[20] Although each of these enterprises was now entirely separate, nothing precluded the original Florentine partners from also working with and for each other from time to time.

Roger Sherman made other promotional pieces for Smith College and Mount Holyoke College, for example, while also joining forces with Larry Hott to coproduce and codirect *The Garden of Eden* (1983), an Academy Award–nominated documentary short, underwritten by The Nature Conservatory. This 28-minute film illustrates the importance of preserving natural habitats for both economic and ecological reasons. Terry Hopkins, who dropped out of Florentine Films before its new incorporation, served as the director of cinematography on *The Garden of Eden,* a role he next shared with Ken Burns on *The Shakers.* Roger Sherman, too, was a production associate, and Buddy Squires a consultant on Burns and Stechler's developing documentary effort. The couple was now starting anew by coproducing *The Shakers* together. They were additionally co-owners of their own small family business, Florentine Films/American Documentaries, Inc., and according to Stechler, "genuine partners" on a film that gracefully captures the depth, beauty, and serenity of its subject.[21]

The history of the Shakers is a uniquely American story that begins with Mother Ann Lee, a former factory worker and illiterate English immigrant, who comes to upstate New York on the eve of the Revolutionary War and pioneers a utopian religious movement, radical in its feminism, pacifism, and communal lifestyle, eventually reaching 6,000 members in 19 communities during its apex in the 1840s. "They called themselves the United Society of Believers in Christ's Second Appearing," announces narrator David McCullough 90 seconds into the documentary over a montage of 13 period photographs of these breakaway Quakers, "but because of their ecstatic dancing the world called them Shakers.

They were ordinary people who gave up everything—homes, families, and livelihoods—to put into practice what they called authentic Christianity." As philosopher Jacob Needleman summarizes toward the end of the film's first chapter, "Opening the Gospel 1770": "America is the land of zero. We start from nothing. That's the ideal of America. We start only from our own reason, our own longing, our own search."

Ken Burns and Amy Stechler Burns's joint exploration into the legacy of the Shakers contains not only the history of the society's rise and decline over two centuries but also onscreen commentaries by three Shaker eldresses and many shots of the sect's material remains, such as its elegant and unadorned architecture and furniture. The 59-minute circular structure of the film commences with a short series of nineteenth-century sepia close-ups of individual Shaker men and women, leading to a misty live silhouette of Hancock's curved stone barn taken at dawn, much as the Burnses had first seen the building early on that August morning in 1981. The soundtrack starts quietly with birds chirping, fading up slowly as a young man sings a well-known Shaker hymn a cappella: "I will bow and be simple, I will bow and be free, I will bow and be humble, Yea bow like the willow tree."

Nearly an hour later the documentary ends almost as simply. "In 1961 Elder Delmer Wilson of Sabbathday Lake [Maine] died. He was the last covenanted male Shaker, and shortly after his death the surviving eldresses decided to close Shakerism to new members forever."[22] Earlier David McCullough had reported offscreen that "today less than a dozen Shakers live in two 200-year-old villages in Maine and [Canterbury] New Hampshire. These are the last Shakers." Sister Mildred Barker of Sabbathday Lake and Eldresses Bertha Lindsay and Gertrude Soule of Canterbury then provide personal reminiscences throughout the film. In the final few minutes, each of these women offers a concluding testimonial to the "love and the influence" of Shakerism. For instance, Eldress Soule notes: "I'm very proud of my heritage and I'm so happy to share what we've had and what we've experienced with all who come in our presence." Fellow Eldress Lindsay follows her statement with a song entitled, "Oh My Sweet Shaker Home," over a radiant midday shot of the stone circular barn, one last time, as the image gently fades away, leaving a closing graphic which dedicates *The Shakers* to the Burnses first daughter, Sarah.

Ken Burns was actually revisiting familiar territory at Hancock Shaker Village, having based his Hampshire College thesis film, *Working in Rural New England*, on another nineteenth-century agrarian preserve—the Sturbridge Village living history museum—located approximately 75 miles southeast in central Massachusetts. The level of technical skill evident in *The Shakers* by comparison, however, reflects just how far he had developed as a producer-director during the eight years between the release of his last college film in 1976 and the completion of his second major professional effort. Burns's working relationship with Geoff Ward, who had just finished a seven-year stint as managing editor (1977-1979) and then editor (1979-1982) of *American Heritage* magazine, began in 1983 when the filmmaker invited the writer to assess a rough cut of the documentary from a historical perspective. "I was asked to review it as a consultant," Ward recalls, "It was the easiest money I ever earned; my role consisted entirely of watching and saying 'wow.'" He observes further that *The Shakers* "seems to me a model of its kind, often beautiful but never romanticized, historically accurate, and genuinely entertaining. It is less a history than an *evocation* of the Shakers."[23]

Ken Burns's stated aim in coproducing and codirecting this film with his wife was a "distillation of history."[24] His increasingly distinctive style was now emerging as far less didactic and informationally oriented than was typical of most historical documentaries at the time. To be sure, *The Shakers* reflects painstaking research, featuring its share of facts, figures, expert interviews, archival photographs, and period music, all "welded together by narration that can soar to near religious inspiration," according to *Time*'s television critic Richard Zoglin.[25] *The Shakers* most importantly, though, "appeal[s] as much to the audience's emotions as to its intellect," reported Leslie Garisto of the *New York Times*.[26] "The resonances of particular lives or events seem to spark powerful emotions within me," relates Burns in explaining his genuine passion for history, "I'm still surprised continually at the depth of my real love, and I think that's the word, of the American past."[27]

Ken Burns, in full collaboration with his wife, continued to sharpen his abilities as a popular historian at this early stage of his career. Together they utilized the film medium as a way of exploring those aspects of the Shakers that were most immediate and relevant to them in the present. Two minutes and 30 seconds into the documentary, for example, the Burnses insert a commentary by historian Ed Nickles, which basically frames how they would like subsequent viewers to think about their topic: "They are in many ways the American Dream. They encompass all the ideas of America—its individualism as well as its collectivism—and . . . they experiment with [them], and they are very successful and yet they are not successful. In terms of lasting, they are perhaps not successful; in terms of what they contribute, they will continually be successful in my view."

The filmmakers then divide their plot structure into nine sequences, with five of the chapters devoted primarily to a historical accounting (i.e.,

"Opening the Gospel 1770," "Gathering into Order 1785," "The Kentucky Revival 1800," "Renewing the Faith 1840," and "Decline"), while the four remaining sections incrementally reveal who the Shakers were, how they lived, and what they believed in from a human interest standpoint. Chapter three, "The World's People," for instance, provides a wide assortment of reactions from many non-Shakers, underscoring the fact that even several significant national and international figures were paying close attention to the activities of this religious and communal sect: "'If their principles are maintained and sustained by a practical life, it is destined to eventually overthrow all other religions,' Thomas Jefferson, 1808"; "'The first people in America, and actually the world, to create a society on the basis of common property are the so-called Shakers,' Friederich Engels, 1845."

Chapter four, "Dance," additionally illustrates through sketches, songs, and narration the rituals and rationale behind the Shakers' "unconventional worship," a place where "worldliness" and "carnality were burned away," making "pure contact with God and each other" possible. Chapter five, "Hearts to God," further clarifies a religion that was, first and foremost, a lifestyle based on "disciplin[ing] the body" as a way of "feed[ing] the spirit." Even though the United Society of Believers in Christ's Second Appearing were strictly celibate, David McCullough notes in a voice-over, "theirs is one of the most enduring religious experiments in American history." Jacob Needleman, moreover, suggests onscreen that much of their creative energy "came from the authentic sublimation of sexual energy." The Burnses then incorporate a series of live stationary shots of chairs, hallways, and stairwells as visual symbols of the Shakers' industriousness, piety, and virtue in the belief that "a true understanding of Shakerism is available, in direct proportion to one's observations."[28] As noted before, the central architectural metaphor in the documentary is the round stone barn in Hancock, described by Amy Stechler Burns in the companion text to the film as "an American cathedral."[29] The Burnses, too, began referring to their work space in Walpole as "the cathedral," reflecting the degree to which they were now identifying with and internalizing the spirit and philosophy of their newest subject.[30]

The progressive nature of the Shaker movement explains in part why this topic appealed so strongly to two young and idealistic filmmakers who came of age during the 1960s and had just moved to a small town in north central New England to pursue their life's work together. "Seventy-five years before the emancipation of the slaves and 150 years before women began voting in America," recounts narrator David McCullough, "the Shakers were practicing social, sexual, economic, and spiritual equality for all members." Chapter seven,

"Reaping the Harvest," portrays the surviving fruits of the Shaker belief that God dwelt in the craftsmanship of everyday work—an inspiring regeneration of the Protestant work ethic—by featuring more of Ken Burns's own footage of the sect's sleek and refined furniture and dwellings. In a later interview, he similarly intimates how the historical example of the Shakers still stays with him, nearly two decades after completing the film: "I don't necessarily spend each day pursuing the spiritual dimension per se. I know that if you work humbly . . . just do the work, these things will come in."[31]

As with *Brooklyn Bridge*, Ken Burns and Amy Stechler Burns approached *The Shakers* as if it was a "labor of a lifetime."[32] By any measure, this historical documentary was a low concept project, pursued by the Burnses out of their shared love and interest in the material rather than from any unreal hopes they harbored of a career breakthrough with the topic. The design and feel of *The Shakers* effectively captures the smallness and delicacy of its subject. The ensuing publication by the same name, *The Shakers—Hands to Work, Hearts to God,* is artfully and lovingly constructed as well, not the usual picture book released on the heels of yet another PBS special. The Burnses, not surprisingly, waited almost a year before *The Shakers* was finally scheduled for its national public television premiere on August 7, 1985. By that time, Ken Burns was already nearing the finish of two more productions, *The Statue of Liberty* and *Huey Long. The Shakers* did garner many positive reviews and eventually won two national awards, a 1984 CINE Golden Eagle and a Blue Ribbon at the 1985 American Film Festival. John Corry of the *New York Times* also selected the film as One of the Year's Best of 1985.[33] Ken Burns had now served as the driving force behind two highly accomplished and well-received hour-long historical documentaries since the formation of Florentine Films a decade earlier. He recognized clearly, however, that he needed to increase his filmic output if he was ever going to establish his own credentials and that of his company on a national level. Over the next five years, therefore, he produced and directed four more feature-length historical documentaries, thus catapulting both his own fortunes and those of Florentine Films/American Documentaries into the forefront of independent nonfiction filmmaking in the United States by the end of the decade.

PRODUCING HIS OWN SPECIAL NICHE

It's interesting that for so long documentary filmmaking has been seen as a lower rung on a career ladder, but I realized that I am absolutely, particularly pleased with the drama that is more dramatic than anything the imagination can think of. I could do ten lifetimes of these stories if I ever started doing them all. It would take me 1,000 years just to cover the last 150 in American history.

—*Ken Burns, 1996*[34]

The mid-1980s were a period of intense creative activity for Ken Burns. He and his wife Amy were able to raise the money for *The Shakers* in a relatively shorter time frame than it had taken for *Brooklyn Bridge,* especially after the latter received an Academy Award nomination for Best Documentary Feature in February 1982. Funding for *The Shakers,* nevertheless, needed to be patiently pieced together from a variety of national and state granting agencies.[35] Attracting financial support is always difficult for independent filmmakers, and as a result, Burns makes a point of not having "a lot of irons in the fire. I concentrate on a few projects," he explains, "two or three things I will definitely do, as opposed to nine or ten things out there in the hopes that one comes in."[36]

While editing *The Shakers,* Burns started thinking about other projects that he might want to develop. One such opportunity was actually suggested to him when he attended the Organization of American Historians annual conference in April 1982 to accept the first ever Erik Barnouw Prize for the year's outstanding documentary film in American history for *Brooklyn Bridge.* David Culbert, a Louisiana State University history professor and chair of the prize selection committee, was in the early stages of researching and writing grant proposals for a possible nonfiction film of his own tentatively titled *Every Man a King: Huey Long and the Rise of Media Politics.*[37] Culbert discussed his nascent idea with the visiting producer-director who was intrigued on first impression and decided to further consider the offer of heading up this particular effort over the coming year. At the same time, Burns was already having discussions with national programming executives at Thirteen/WNET about directing and coproducing, with his old friend and associate Buddy Squires, a follow-up to *Brooklyn Bridge* on an even more celebrated and internationally prominent New York City landmark—the Statue of Liberty.

Ken Burns, consequently, was making preparatory production plans along two parallel tracks by the end of 1982 and into the next year, gradually committing more and more of his attention toward exploring a pair of new historical topics, especially since postproduction on *The Shakers* was well along during the fall of 1983. "I'm aware of a subject, a story, an individual," Burns reveals, "and there might be a kind of intellectual interest in that and maybe even the thought that this might someday make a good film. But usually it is a combination of very powerful, emotional resonances that convince me to finally make the film, and when the bell goes off inside, then there is really no stopping."[38] He was now hitting the ground running on two separate fronts as research began in earnest on *The Statue of Liberty* in October and on *Huey Long* in December of 1983.

The creative team on *The Shakers,* by and large, were people whom Burns knew from Hampshire College and Florentine Films—Amy Stechler Burns, Terry Hopkins, Roger Sherman, and Buddy Squires—or individuals who had advised and worked with him on *Brooklyn Bridge,* most specifically, narrator David McCullough and co-writers and researchers Wendy Burns Tilghman and Tom Lewis.[39] On his next few projects, Burns substantially widened his net of collaborators. Through McCullough and Geoff Ward, the producer-director was now becoming fast friends with the circle of talented popular historians who either edited or contributed articles to *American Heritage.* Richard Snow, for example, who was then managing editor at the magazine and had worked at the publication since his Columbia College days in the late 1960s, was especially helpful in accessing relevant stills from the vast photo archives at the American Heritage Publishing Company. These pictures, most of which were already in the public domain and free for Burns's use, provided him with a beginning framework of imagery for both *The Statue of Liberty* and *Huey Long* (as well as a future stockpile of Civil War photographs, amassed over the years at the behest of *American Heritage's* founding editor Bruce Catton, the celebrated historian who during the early 1960s had published the best-selling *Centennial* trilogy on the conflict).[40]

Upcoming anniversaries were already an important part of the popular appeal of Ken Burns's two new projects: the Statue of Liberty was soon approaching the centennial of its first unveiling in New York harbor on October 28, 1886, and the assassination of Huey Long at the Louisiana State Capitol in Baton Rouge on September 8, 1935 was nearing its 50-year memorial. Each of these milestones provided an added urgency to these projects once they commenced, as prolific writer and contributing editor to *American Heritage* Bernard Weisberger began drafting the initial script for *The Statue of Liberty* and Geoff Ward made his first foray into screenwriting with *Huey Long.* As Ward describes the experience,

> one of the things that Ken has that I think is an enormous advantage over a
> lot of other filmmakers who have made documentaries in the history field, is
> that he genuinely believes history is absolutely fascinating. A lot of people who
> try their hands at documentaries have to convince themselves as they go along
> . . . [but] Ken from the first was completely convinced that you could make
> something terribly compelling and emotional and powerful out of archival

material. Since that's something I always believed in as editor of *American Heritage* and writer and so on, we really sort of hit it off. And he asked me to write a film on Huey Long, which I did. I had never written a film. I had never read a film script that I can remember, but I thought I'd try, and I did *Huey Long* and we had a wonderful time working on it together. We've been working together off and on ever since.[41]

Burns truly found kindred spirits to interact and consult with at *American Heritage*. They too were accomplished popular historians who shared the producer-director's unabashed enthusiasm and love of history. Burns often calls himself an "amateur historian," claiming that "I don't want to be ashamed of that. I think quite often the best histories are done by amateurs who have a more direct connection to the visceral power of history, who often have the tools to be able to communicate it to large groups of people."[42] Burns's characterization of himself as an amateur makes sense in the late eighteenth- and nineteenth-century meaning of the term, someone who pursues an art, science, or some other endeavor as a pastime rather than a profession. Thomas Jefferson, from this vantage point, was considered an amateur architect "only insofar as [he] did not earn a living designing buildings; although self-taught, [his] expertise was professional" by modern standards.[43]

Ken Burns is, likewise, an able if "self-taught" historian in his own right, but he is not a *professional*. In contemporary America, the term *professional* suggests a person who has made a lifetime commitment to a specialized career and thus belongs to an exclusive and highly select group. A *professional historian*, in this way, is a scholar who belongs to the academy. An *amateur*, in contemporary terms and by contrast, is not to be taken all that seriously; he or she is considered a beginner, a dabbler, or in the worst case scenario, a dilettante. "I just wanted to say that I wasn't an historian in the traditional, professional sense," admits Burns, "and I think it may have been a little insulation or armor that would protect me."[44] In today's parlance, therefore, he is more precisely a *popular historian* than an *amateur*, who uses the power and influence of film to reach well beyond a scholarly audience with his television histories. What he shared immediately with the people at *American Heritage* was a similar attitude and mission. As Richard Snow described in a 1985 letter to Burns: "I am, both by training and inclination, involved in the same thing you are: making the past accessible to a lay audience."[45]

The Statue of Liberty, specifically, provided Burns with a topic that possessed a much larger built-in audience than either *Brooklyn Bridge* or *The Shakers*. His creative challenge as a professional filmmaker and popular historian was "to take a subject that's been beaten to death—cliché-ridden, misused and overused—and pull something out of it."[46] He approached *The Statue of Liberty* as both a historical rendering and a cultural reinterpretation of this unique architectural symbol from a new generational perspective. Like *Brooklyn Bridge,* the producer-director's newest work developed into an appreciation of yet another unique American landmark whose history and symbolic importance are now part of the national experience. Burns's resulting 58-minute film stands as a subtler alternative to all the commercialized hype that surrounded the statue's restoration and rededication during the summer and fall of 1986. In contrast, David Wolper, arguably television's most successful independent documentary producer of the World War II generation, was executive producer of *Liberty Weekend* for ABC. This star-studded extravaganza, complete with 300 tall ships, the Boston Pops, and the greatest fireworks display in American history up to that time, attracted 1.5 billion viewers worldwide during the four-day telecast in October 1986.[47] Burns's effort was comparatively far more modest and even-handed, "establish[ing] a central thematic tension" according to his original two-page treatment for the film, "between liberty as a finished accomplishment . . . and liberty as an unfinished potential."[48]

The Statue of Liberty was modeled on the two-part structure which had worked so well for *Brooklyn Bridge* just three years earlier. The first half of the historical documentary, entitled "The Idea," traces the conception, construction, and installation of the monument. This 29-minute and 10-second section follows the 21-year odyssey of sculptor Frédéric Auguste Bartholdi, who set in motion Professor Édouard Laboulaye's original "idea" in 1865 to design, build, and then present to the citizens of the United States a monumental statue from the French people that would commemorate the centennial of the Declaration of Independence. The story of Bartholdi's artistic obsession, which sported the face of his mother and the body of his mistress, is a compelling biographical history spanning both sides of the Atlantic. It furnishes an abundance of celebrity cameos, such as engineer Alexandre Gustave Eiffel and *New York World* publisher Joseph Pulitzer and climaxes with the 1886 dedication of the Statue of Liberty in New York harbor, ten years later than originally planned.

The second half of *The Statue of Liberty,* entitled "The Promise," explores the enduring impact and significance of the monument in contemporary American life. This part, lasting 28 minutes and 50 seconds, contains an array of commentaries from well-known and ordinary people alike about the symbolic meaning that the statue holds for them, interspersed with numerous examples from popular culture where the landmark has been incorporated into a wide variety of movies, advertisements, and political cartoons. Ken Burns's stated goal in this section is to utilize "these voices and interviews . . . not so much to answer as to ask three main kinds of questions: What is liberty? What threatens liberty? Why should we care about the Statue of Liberty?"[49] Approximately 15 minutes from the end of *The Statue of Liberty*, a succinctly structured four-minute scene highlights several of the more revealing responses Burns received to these three basic questions while conducting his interviews for the documentary.

Narrator David McCullough begins this editing cluster by proclaiming over a live static close-up of the monument that "the Statue of Liberty is an act of faith," followed by an onscreen commentary by statesman Sol Linowitz recalling his parents' arrival "from overseas" and their outlook as immigrants: "They came enchanted by an idea that here they could be themselves and be the best that was in them to be. They were coming to a place where every human being could stand erect and with dignity as a child of God. That's what I think they believe in—and that's what I see in it. It reminds me of how much we yet have to do to achieve the full promise of that statue." The next shot is a 25-second excerpt from Martin Luther King's "I Have a Dream" speech, climaxing with a critique by writer James Baldwin which ends, "For black Americans, for black inhabitants of this country, the Statue of Liberty is simply a very bitter joke, meaning nothing to us." Filmmaker Milos Forman, an expatriate from Czechoslovakia, then concludes this scene with an offscreen commentary over a majestic long shot of the statue silhouetted against a luminous orange sky: "There is something deep down in this society which is so valuable, so strong, and so inspiring that in spite of all the difficulties, all the troubles, all of evil, all of the crime you see around yourself, all of greed, all of hypocrisy you see, that still this is the best. And that lady up there symbolizes that."

In this brief but revealing scene, Burns and his editor, Paul Barnes, present complimentary and critical reactions to both the Statue of Liberty and the amount of freedom that it actually embodies. Martin Luther King's and James Baldwin's statements, in particular, call into question all the celebratory good will expressed at the outset by McCullough and Linowitz, although the positioning of Forman's remarks right after Baldwin's heartfelt reference to racial prejudice and injustice clearly softens the author's frank characterization of the monument as a "bitter joke." This sequencing of shots, most importantly, captures Ken Burns's liberal pluralist outlook in action. The filmmaker's overall portrayal of America as represented in *The Statue of Liberty* is of a nation coming together around its common democratic ideals, despite its still unfulfilled promise of liberty for all. The filmmaker's majoritarian aspirations are reinforced one last time in the film's final commentary, delivered once again by narrator David McCullough, only this time onscreen:

Liberty is what we Americans have always wanted first of all. It's what the country was founded for. It's what the Revolution was fought for. All the great songs and sayings and pronouncements of those Revolutionary figures were about liberty. And they knew what it meant [cutting to a close-up of McCullough]. And because the French sent the statue here, it was their way of saying implicitly—we recognize that [the Statue of Liberty] is a gateway to a new world and to the hope of the world [fade out to credits and the closing musical theme].

Paul Simon's bittersweet 1973 anthem "American Tune" both opens and ends the film, reflecting Ken Burns's generational sensibilities and his preferences as a director. The overall tone of *The Statue of Liberty* is similarly one of underlying disillusionment, palpable beneath the more patriotic sentiments which are clearly a vital part of this statue's symbolic legacy. The opening 2-minute 20-second prologue is a case in point. The documentary starts with a 20-second epic long shot of a steely gray Statue of Liberty framed against a brilliant pink sky. Offscreen the voice of actor Paul Roebling reads a slightly adapted version of Thomas Jefferson's historic introduction to the Declaration of Independence: "Listen—we hold these truths to be self-evident that all men are created equal, that they are endowed by their creator with certain unalienable rights, that among these are life, liberty, and the pursuit of happiness." The screen then fades to black for five seconds before "American Tune" slowly fades up on the soundtrack as an understated counterpoint to Jefferson's inspirational words.

The melody of "American Tune," adapted by Simon from a Bach chorale, produces a lovely if melancholy texture to the rest of the scene. Six leisurely close-ups of the statue comprise the next minute, one of which focuses for 12 seconds on the tablet that Lady Liberty is holding with the date July 4, 1776, etched on it in Roman numerals. The audience next shares one particular immigrant's point of view, as a rephotographed image gradually zooms out over 17 seconds, revealing a man gesturing toward the statue as his whole family looks on with all their belongings lying on the dock at New York harbor. It is an intimate and touching moment as Paul Simon can be heard singing softly in the background: "I don't know a soul who's not been battered, I don't have a friend who feels at ease. I don't know a dream that's not been shattered, or driven to its knees." A montage of eight more newcomers from overseas are shown in rapid succession over Simon's rendition of "American Tune" before the scene finally ends with two long shots of the statue, each lasting seven seconds, as the word *Liberty* first appears by itself on the screen for a moment before dissolving seamlessly into the entire title, *The Statue of Liberty*. This introduction is wonderfully realized, emotionally arresting, and suggestive of both the filmmaker's deep ambivalence about the present as well as his more urgent need and desire to believe in the monument's promise.

The Statue of Liberty was the first of many joint collaborations between Ken Burns and Paul Barnes, another baby boomer who is two years older than the producer-director. Barnes was already recognized as one of the finest young nonfiction film editors in the country, having just received the rare honor of being nominated the previous year for two Eddie awards in the Best Edited Documentary category by the American Cinema Editors Guild. One of these nominations was for his work on the vibrant ode to Gospel music *Say Amen Somebody* (1982), directed by George Nierenberg. Barnes won the Eddie for *Wasn't That a Time!* (1982), director Jim Brown's chronicle of The Weavers, a popular folk music group from the late 1940s and early 1950s, which culminates with a rousing final reunion concert at Carnegie Hall in 1980 shot mainly in cinema verité style. *The Statue of Liberty*, in comparison, projects a much slower and more reflective tempo, indicative of a film director who also happens to be a committed still photographer. As Barnes recalls,

> When I first began to work with Ken on *The Statue of Liberty*, I was cutting some of the still photograph sequences and I was tending, having not worked on a film with still photographs before, I was tending to cut them a little bit too quickly. And, he looked at a couple of scenes I did, he said to me, "look you're using too many shots and you're not holding them long enough. I want people to feel as if they can live in this photograph, allow the photograph to breathe, allow the audience to live in it, allow them to explore it with their eyes."[50]

With each new assignment, Burns and his creative associates were learning how best to integrate the seemingly antithetical traditions of the social documentary, which the producer-director had learned and embraced at Hampshire College, with the more popular storytelling techniques which were the forte of those Hollywood directors, especially John Ford, whom Burns grew up admiring on late night TV. Made-for-television history is as much about "myth . . . as fact," Burns was increasingly realizing with each new production, "because myth tells you much more than fact about a people."[51] The Statue of Liberty is an inherently mythic topic with all of its broader ties to our historic relationship with France dating back to the Revolution, its primary aim of commemorating the Declaration of Independence, and its symbolic linkage to the nation's continuing legacy of immigration. The film, too, provides a discernable fact-filled account of the monument's conception and building, which is depicted carefully and credibly during the first half of the documentary.

The other major project that Ken Burns was preoccupied with from late 1983 through 1985 was again a highly combustible mixture of history and myth. *Huey Long* rivaled the larger-than-life plot lines and characters of Hollywood. Never before had Burns chosen a topic that was so intrinsically dramatic, even though *Brooklyn Bridge, The Shakers,* and *The Statue of Liberty* certainly have well-told stories as a basic part of their structures. Veteran producer, director, and screenwriter Robert Rossen had actually garnered a Best Picture Oscar in 1949 for his adaptation of Robert Penn Warren's Pulitzer Prize–winning novel, *All the King's Men,* based loosely on the real-life experiences of Huey "Kingfish" Long of Louisiana.[52] This celebrated Columbia studio production, however, pales in comparison to Burns's nonfiction biography, complete with its expertly crafted narrative, a complex multifaceted rendering of its main character, and Long's own spellbinding personality making periodic cameos courtesy of stock footage, archival photographs, and old radio recordings.

Director Ken Burns coproduced *Huey Long* with Richard Kilberg, whom he had originally met at Thirteen/WNET while making *Brooklyn Bridge.* Kilberg was then an assistant director in the national programming department at the station, before moving briefly to HBO as a production supervisor in 1982. The next year, Kilberg became a principal partner in RKB Productions, based in New York, which cosponsored *Huey Long* along with Florentine Films/American Documentaries and WETA-TV in Washington, D.C. Ward Chamberlin, then president at WETA, became another mentor to Burns, someone whose profes-sional example and unprecedented support cemented the young filmmaker's continuing and mutually beneficial relationship with that public TV station from late 1983 onward.[53] As Ken Burns remembers,

> [Chamberlin] partially funded *Huey Long,* and one of the most flabbergasting moments in my life was when I was short $50,000 on *Huey Long,* and was merely hoping that WETA would help me raise the money, and he said well I don't see why we can't do that. And for the first time just about ever, somebody put cash on the barrelhead and I walked out of the meeting with a check for $20,000, and he was apologetic that he couldn't give me the whole thing. I had never seen that in public broadcasting or television production ever, and I haven't seen it again except by him, and he reminded me what it's all about. It isn't about perks, it isn't about carpeting on your floor, it's about the quality

of the programming, and he was willing to literally put his money where his mouth was.[54]

Ken Burns's web of personal and professional contacts now extended well into the public television system with both past and current production personnel and executives, such as Kilberg and Chamberlin. Geoff Ward, from the circle of associates whom Burns knew through *American Heritage,* completed his first version of the *Huey Long* script on July 4, 1984. In close consultation with Burns over the next five weeks, Ward focused expressly on turning "straight historical reporting [in]to scenes," with two more drafts of his narration mailed out for feedback on August 14 to the four main scholarly advisors on the film—Alan Brinkley, William Leuchtenburg, Arthur Schlesinger, Jr., and Robert Snyder.[55] These consultants, along with David Culbert, who was now an associate producer for research on the film, responded encouragingly to the way the documentary was progressing. Leuchtenburg, for instance, wrote back on September 19, "I am absolutely enthusiastic about it . . . I am often asked to read scripts like this, and I think the narrative is by far the best I've ever read."[56] Ward was clearly demonstrating a natural affinity for scripting, as Burns asked him to collaborate with Bernie Weisberger on *The Statue of Liberty* during the remainder of the fall.

Ken Burns, furthermore, renewed his many working relationships with his Florentine friends and family members during the course of these two projects. Amy Stechler Burns edited *Huey Long* and helped out with supplemental writing on *The Statue of Liberty.* Buddy Squires, who coproduced *The Statue of Liberty,* was also postproduction supervisor on *Huey Long* and shared camera duties with Burns on both projects. Roger Sherman was production manager on *The Statue of Liberty* and a sound technician on *Huey Long.* Burns even invited his brother, Ric, to spend a short hiatus away from his doctoral studies in English and comparative literature at Columbia University to oversee the scripting process of *Huey Long,* where he contributed some of his own writing after being a production assistant and editing consultant on *The Statue of Liberty.* Ric's brief interlude from academe became permanent, as he assumed an even greater role as his brother's coproducer on the developing Civil War Film Project in early 1985. *The Civil War* soon received $20,000 in developmental monies from WETA and another $20,000 in startup funds from PBS. "This $40,000 not only fed Burns and his family for all of 1985, but was used to pay consultants, fund the writing of grant proposals and set up an office for the production."[57] Easily the most versatile and driven Florentine family member of all, Ken Burns was, additionally, supervising the editing of *The Statue of Liberty* and *Huey Long,* bringing these two projects to closure in just over two years.

Huey Long was, significantly, Burns's first full-fledged historical biography. In each of his three major quasi-biographical documentaries so far, he had utilized first-person stories and anecdotes, the stuff of personal heroism and tragedy, from inside the framework of larger historical currents and events. In his newest film, however, the biographical approach became even more pronounced with a protagonist who virtually demanded the extra attention. Long, for instance, once told reporters, "I am *sui generis*, just leave it at that."[58] He was, indeed, a genuine "American original." Robert Penn Warren, whose tenure as an English professor at Louisiana State University (1934-1942) overlapped the time that the Kingfish ruled the state (1928-1935), sets the proper tone and context, just 70 seconds into *Huey Long*, with one of his many commentaries and readings throughout the film, delivered in his own distinctive rolling Kentucky cadence over a graphic containing the same exact wording: "If you were living in Louisiana you knew you were living in history defining itself before your eyes and you knew that you were not seeing a half-drunk hick buffoon performing an old routine, but witnessing a drama which was a version of the world's drama and the drama of history too: the old drama of power and ethics." Ken Burns likewise characterized his latest documentary as "a tragic almost Shakespearean story of a man who started off good, went bad, and got killed for it."[59] *Huey Long* held a special resonance for the filmmaker, marking yet another subject which fit his generational view of life and politics forged during the 1960s and by the formative experience of Watergate.

The overall 88-minute narrative structure of *Huey Long* is roughly divided into thirds. The first section is composed of a short prologue and two chapters, "Every Man a King" and "The Kingfish," and essentially follows Long from his middle-class upbringing in the northern Louisiana township of Winfield through his meteoric rise in law and politics to the governorship at the ripe old age of 34. The middle portion, comprised of "Sure Enough Bigshot" and "Share the Wealth," covers his concurrent terms as governor and U.S. senator, chronicling his abundant political skills and the paradoxical nature of his leadership: Long was simultaneously a champion of the poor and powerless, ushering a Louisiana ravished by poverty, illiteracy, and disease into the twentieth century, while also amassing unprecedented amounts of wealth and power himself. This second third of the documentary ends with Long's bold emergence on the national scene, his becoming a legitimate political rival to F. D. R. during a depression that "had transformed the whole nation into something like back country Louisiana," reports narrator David McCullough.

The concluding 25-minute sequence of the documentary, entitled "A Wildness in the Air," deftly brings the electrifying life of Huey Long to its untimely and premature end. With the citizens of Louisiana on the brink of civil insurrection, bitterly divided over the Kingfish's undemocratic uses of authority and control, Long is gunned down in the state Capitol by the most unlikely of assassins, a young surgeon named Carl Weiss, Jr., who also happens to be the son-in-law of a political foe. Justice John Fournet, a former legislator and Long supporter, delivers an eyewitness account of the shooting, as Burns creates an intensely dramatic scene by mixing narration with archival photographs, a pen-and-ink sketch, old black-and-white newsreel footage of the aftermath, and live contemporary color shots taken at the scene of the crime. Characteristic of the film as a whole, the richly enigmatic nature of Long is reflected in a 5-minute 28-second summation to his killing as a dozen different people who both loved and hated him offer an array of varying opinions.

A cross section of examples from this editing cluster includes journalist Betty Werlein Carter, whose husband was newspaperman Hodding Carter II ("You didn't know who it was. It could have been anybody. And I hate to say, we really hoped he would die. Now that's a terrible thing to say."); Huey's sister ("And his last words were 'don't let me die, I have so much to do.'"); his son Russell ("It's just something I had to live with."); journalist I. F. Stone ("I was very impressed with him, but it's a terrible thing to say, I was really glad when they shot him. I don't believe in terrorism or assassination, but he could have become an American dictator."); a political associate ("I lost one of the best friends I ever had."); and Thomas Weiss, brother of the assassin ("A horrible thing occurred, and great sorrow in both the Long and Weiss families for what went on, and the only thing we could do was pray for both of their souls."). Cecil Morgan, who led a group of state legislators who impeached but failed to convict Long at the time, offers a much broader evaluation of Long's legacy just one minute later: "I think it was necessary for the state, somebody with his qualities to come forward, and I think he muffed it. I think he had the capacities for greatness, and I think he did some things that stimulated the state enormously, and that's on the good side. On the bad side, he left us with a heritage [of political corruption and patronage] that we have not yet recovered [from]."

Burns, appropriately, leaves the last words to Robert Penn Warren, reading again from *All the King's Men:* "What happened to his greatness is not the question [a black screen fades up slowly to a head shot of Huey Long as a young boy]. Perhaps he spilled it on the ground the way you spill a liquid when the bottle breaks [dissolve to another photograph of Long as a teenager wearing a skimmer]. Perhaps he piled up his greatness and burnt it in one great blaze in the dark like a bonfire [dissolve to a still of Long as a young man in his early 30s], but then there wasn't anything but the dark and the embers winking [dissolve to a final picture of Long right before he was shot to death at 42]. Perhaps he could not tell his greatness from ungreatness and so mixed them together that what was unadulterated was lost [cut to close up of Warren reading]. But he had it. I must believe that [fade to black as credits roll over Randy Newman's soulful and sardonic 1974 song, 'Louisiana 1927']."[60] In the end, Burns skillfully blends fiction with fact by closing his film with a selection from Warren's poetic novel, spoken by the writer himself who is, after all, a living witness to history. The audience, in turn, is invited to reflect upon (and experience) the memory of Huey Long as he somehow grows old one last time in four period photographs—his life literally passing before each viewer's eyes— in a powerfully moving coda replete with mythically charged suppositions about his own final undoing and his bona fide greatness.

Ken Burns was already an experienced producer-director with a rapidly growing list of finely crafted documentaries when at age 32 he completed *Huey Long*. By this time, he had more than confirmed his considerable creative and supervisory abilities in practically every aspect of filmmaking. Still, the one stylistic feature that peers and audiences alike most associated with him was his rephotographing technique.[61] Susan Sontag writes in *On Photography* that a "photograph is only a fragment and with the passage of time its moorings come unstuck. It drifts away into a soft abstract pastness, open to any kind of reading (or matching to other photographs)."[62] Burns fully understands this dynamic and utilizes it to both inspire his own imagination and add emotional depth and texture to his storytelling. "We film old photographs," he explains, "with an energetic and exploring camera eye, sort of not content to film it at arm's length and take back merely its static presence, but to go in to look and investigate the new stories that are within the photographs, and I think more than anything, to listen to what the old photographs have to tell us."[63]

Approximately 12 minutes into chapter one, "Every Man a King," for instance, a brief 55-second vignette containing five period stills ideally captures the subtle effectiveness of Burns's poetic approach to history. This section is part of a slightly larger two-minute scene that begins with stock footage of Long, speaking energetically to a large crowd, as David McCullough recounts offscreen: "His listeners loved to hear him lash the rich and powerful—the thieves, bugs, and lice who dared oppose him." The screen, then, cuts after this 45-second long shot to a static image of the Kingfish with his arms and hands spread outward, speaking into an old NBC microphone at another public gathering, as the lens slowly tilts upward, closely inspecting his standing presence on the podium. This photograph almost instantly springs to life with the clamor of cheering people fading up on the soundtrack several seconds before McCullough begins his narration: "But they loved still more his vision of a new Louisiana." The dramatic *coup de grâce* of this filmic moment occurs with the introduction of Jay Ungar's heartrending lament, "Ashokan Farewell," the very same violin piece that would become the signature theme of *The Civil War* five years later. Burns first and foremost brings these old archival pictures alive by synesthesia, or the process by which one type of sensory stimulation enhances another. In this specific case, the simulated realism of the crowd noises, the narrative context provided by the narration, and the climactic emotional force of the background music literally jump starts this static image of Huey Long in such a way that the audience is better able to suspend its disbelief, thus perceiving the film's protagonist as vital (and maybe even moving for just an instant) in that photograph.

Ken Burns also employs accompanying sights as well as sounds to similarly animate these old archival images. This same scene next cuts to an old black-and-white photo of a majestic oak tree, as the haunting strains of "Ashokan Farewell" and David McCullough's expressive narration continue offscreen: "At the little Cajun town of St. Martinville, he set forth his hopes for the future. 'It is here,' he said, 'under this oak [dissolve to a live color shot tilting slowly downward on the same oak tree as it looks today], where Evangeline waited for her lover who never came. This oak is an immortal spot, made so by Longfellow's poem. But Evangeline is not the only one here in disappointment [dissolve to a black-and-white photo of Myrtle Bigler of Atchafalaya River as a young girl]. Where are the schools that you waited for your children to have [dissolve to live color shot of Myrtle Bigler as an old woman]? Where are the roads and highways [dissolve to black-and-white picture of Edmond Riggs of St. Martinville at about 12 years old] that you sent your money to build [dissolve to live color shot of Edmond Riggs as an old man] that are no nearer now than ever before [dissolve to black-and-white still of Alcide Verret of Atchafalaya River as a young boy]? Where are the institutions to take care of the sick and the disabled [dissolve to live color shot of Alcide Verret as an old man]? Evangeline wept bitter tears in her disappointment [cut to stock footage of a poor middle-aged woman scrubbing clothes at a wash tub outside her small tenant home on the Bayou], but it only lasted through one lifetime. Your tears have lasted through generations. Give me the chance to dry the tears of those that weep here'" [as "Ashokan Farewell" reaches its poignant conclusion offscreen].

In this specific example, Ken Burns revivifies five early twentieth-century black-and-white photographs by merging them sequentially with contemporary color footage. Still and moving images of the actual Evangeline oak and the same three average Louisiana citizens correspond one-to-one, young-to-old over the course of just 55 seconds. Burns recognizes that film and television are incapable of rendering temporal dimensions with much precision. The media have no analogues for the past and future tenses of written language. These visually based media, instead, amplify the present sense of *immediacy* out of all proportion. The producer-director thus capitalizes on this characteristic of film grammar by mixing and matching historical source material, such as these five static photos, with contemporary sights and sounds, putting them all on an equal footing in the present tense. In this particular scene, the live quality of the contemporary color footage encourages audiences to interpret the matching black-and-white stills as if they were also alive and animated. Huey Long's populist message and raw political talents, in this way, grow far more immediate and tangible to current viewers. Audiences can, in effect, experience and better understand the charm of his rhetoric and the appeal of his message for ordinary voters throughout Louisiana at the time—such as the Myrtle Biglers, the Edmond Riggses, and the Alcide Verrets—who McCullough describes in his offscreen narration as the "trappers and fishermen of the Bayou, redneck farmers from the hills, sharecroppers and tenants and small town storekeepers, Catholics and Protestants alike." Instead of overwhelming viewers with too many details, Burns strategically provides just enough information about Huey Long's platform for governor in six representative lines from his stump speech, thus connecting his concerns for education, highways, and health care with memorable imagery—both old and new—as well as with an evocative melody, which together convey to audiences a deeply emotional sense of the past with great staying power and personal impact.

Huey Long premiered in a special screening at the Louisiana State Capitol on September 8, 1985, exactly 50 years to the day after the Kingfish had been fatally shot in the corridors of that building. The documentary was only one of three official American entries in the prestigious New York Film Festival where it was shown on September 28, 1985. Reviewing the motion picture for the *New York Times*, critic Vincent Canby called it a "meticulously researched, graceful, funny and disturbing film."[64] Canby subsequently included it in his Best 10 list of 1986, following the documentary's limited though triumphant theatrical run in selected art house theaters around the country in February and March. *Huey Long* was additionally acclaimed at film festivals throughout the United States and Canada, including Telluride, Hawaii, and Vancouver; featured at the Smithsonian Institution and the Louisiana State University Writers Conference; and awarded Ken Burns's second Erik Barnouw Prize in five years for outstanding historical documentary of 1986 by the Organization of American Historians (OAH). The film, which missed an Academy Award nomination by just two votes in February, received its initial PBS telecast eight months later on October 15, 1986.[65] ABC correspondent Jim Wooten called *Huey Long* "quite possibly the finest political biography committed to film."[66]

It is also one of the foremost documentaries of any kind—historical or otherwise—produced during the 1980s.

The Statue of Liberty was nearly as celebrated and honored as *Huey Long* upon its release, some six weeks after the latter's debut. Burns's second major nonfiction film on a celebrated New York City landmark was first screened as part of an official midweek event at the French Embassy in Washington, D.C., on October 23, 1985. *The Statue of Liberty* premiered on national television the following Monday, October 28, kicking off the centennial festivities on the ninety-ninth anniversary of the monument's inaugural unveiling. Among its many awards, the film was nominated for both an Oscar and an Emmy in 1986. It was, indeed, a worthy follow up to the slightly more satisfying and original *Brooklyn Bridge*, which Burns still describes as "the first historical film of that generation. There had been others, but not in an attempt to be strictly using film as history, as a new kind of visual history."[67]

Ken Burns, in hindsight, had created his own special niche in the independent filmmaking world by the end of 1986. David McCullough, once so wary of collaborating with the young producer-director on *Brooklyn Bridge*, was now coming to believe that he was "by far the finest documentary filmmaker in the country today."[68] Burns and his newly enlarged stock company of creative associates had worked virtually nonstop to produce three films in four years at the recently restructured Florentine Films/American Documentaries. "He's running a mom and pop shop up there," added McCullough at the time, "and doing extremely well against tremendous competition."[69] If anything, Burns's vigorous work schedule was about to get even more strenuous with preparation, research, and preliminary scriptwriting on *The Civil War* already well underway by December 1986.[70] He also decided to concurrently make two additional feature length documentaries—*Thomas Hart Benton* and *The Congress*—works which further reinforced his own distinctive approach, outlook, and position as America's preeminent producer-director of historical nonfiction films.

HONING HIS HISTORICAL VOICE

I must admit that I have, in many ways, made the same film over and over again.
—*Ken Burns, 1996*[71]

Ken Burns was focusing his full attention on *The Civil War* for the first time during the summer months leading up to the successive debuts of *Huey Long* and *The Statue of Liberty* in September and October of 1985. The seed money he had obtained from WETA and PBS's national programming development fund "was basically keeping me alive," Burns discloses in retrospect, "I am an independent filmmaker, and I am precluded by the provisions of the Corp[ora-tion] for Public Broadcasting, the National Endowment for the Humanities and other foundations from having any contingency or profit margin built into my budget, so when a project is over, I'm out of work."[72] He, in turn, "immediately agreed to take . . . on" a new film biography of artist Thomas Hart Benton when he was invited to participate in November 1985 by Peter McGhee, program manager for national productions at WGBH-TV in Boston.[73] Burns instantly responded to the Benton proposal as a subject that fit perfectly into his ever expanding body of work. As he went along, Burns came to recognize more and more how much he actually had in common with the painter. They both shared a total dedication to craft, a desire to reach a broad-based popular audience, and a lifelong love of American history and myth.

Thomas Hart Benton was the brainchild of Henry Adams, a curator of American art at the Nelson-Atkins Museum in Kansas City. Adams's main responsibility at the time was to oversee an upcoming retrospective of Benton's work, the largest ever planned, which was scheduled to open on April 15, 1989, the centennial of the artist's birth. Interestingly, he is also a lineal descendant of John Quincy Adams, the former president and congressman, who furthermore happened to be a sworn political enemy of Missouri's first senator, Thomas Hart Benton (1782-1858), the great-uncle and namesake of the artist. Benton's father also was a congressman, and hoped his son would follow in the family tradition by becoming a lawyer and a politician. The young Thomas Hart Benton (1889-1975), instead, struggled as a painter for years. He eventually made his mark as the country's best-known muralist during the 1930s and early 1940s, even finding himself featured on a December 1934 cover of *Time* magazine as the leader of a new Regionalist movement, along with John Steuart Curry and Grant Wood, who together were bucking the prevailing tide in the art world by advocating American subject matter above all else, especially scenes that extolled the people and culture of their native midwest.

While researching Benton for the exhibition and a related catalogue he was writing, Adams discovered a wealth of film footage, photographs, and audiotape recordings of the artist, which initially gave him the idea of nurturing the creation of an accompanying film on the subject. He approached a number of key programming executives at public television stations in Kansas City, New York, and Boston, and received his most enthusiastic response from Peter McGhee at WGBH who, serendipitously, had known Benton as his neighbor on nearby Martha's Vineyard, where they both had owned summer homes. McGhee agreed to back the proposal in July 1985 on the proviso that the program not be designed as merely an adjunct to the exhibition but as "an independent statement aimed at an enormous national audience." The goal he had in mind was "to produce a film classic, the definitive film treatment of one of America's greatest painters."[74] Adams and McGhee, moreover, discussed a variety of filmmakers who might be appropriate for the project, and they unanimously agreed that Ken Burns was "their first choice as the best possible person to create the documentary." They were particularly impressed with *Brooklyn Bridge* and how it demonstrated his "extraordinary visual sense," his "ability to grasp complex intellectual issues," and his "gift for getting to the emotional heart of his subject matter."[75] Their impressions were reconfirmed again when they saw *Huey Long* in early 1986. *Thomas Hart Benton* was clearly an ideal topic for Burns. Although it provided him with the new challenge of presenting works of art on film, the story of Benton's life also offered the producer-director with yet another custom-made protagonist who fit comfortably into the American original prototype.

All of Ken Burns's television histories up to this point had been populated with seemingly ordinary men and women who rose up from the ranks of the citizenry to become paragons of national (and occasionally international) achievement in engineering, religion, art, or politics, always persisting against great odds. The beginning 2-minute 10-second scene of *Thomas Hart Benton* represents anew this recurring narrative model. The first five shots, lasting a combined 45 seconds, feature three white-on-black typographical statements intercut between two vibrantly colored Benton paintings. The stark contrast between text and imagery is visually arresting, as a solo piano is heard playing a traditional folk tune quietly on the soundtrack. Significantly, Burns presents the basic facts of his topic through graphics rather than voice-over narration as his way of focusing primary attention at the start on what viewers are about to see not hear.[76] The opening lines of text appear in the following order:

1. Thomas Hart Benton painted America.
2. For more than seventy years he painted its cities and small towns, its farms and backwoods. He painted its people, too: faith healers and lovers, politicians and soda jerks, farmers and movie stars.
3. Most of all, Benton wanted to make his art available to ordinary people. To do that, he took on the whole art world and embroiled himself in endless controversy.

The broad parameters of the 86-minute plot structure are thus established from the outset. The remaining 85 seconds of the opening scene are spent isolating ten vignettes from Benton's mural cycle on the popular arts, *The Arts of Life in America* (1932). The sprightly sounds of Dixieland jazz accompany a rapid montage of typical Benton characters: blue collar workers on the job, men watching women dancing in a saloon, a couple by themselves stealing a kiss. Burns's ever-moving camera eye—panning, tilting, zooming—sweeps upward one last time at the end of this section to rest on a dapper self-portrait of the mustachioed artist sitting alone at his easel. The title, *Thomas Hart Benton*, slowly appears on the screen just beneath his chin, as the painted Benton stares back defiantly at the audience, bringing this rousing introduction to its personally revealing conclusion. Thomas Hart Benton was apparently a person-ality who was hard to ignore. As painter Vincent Campanella relates two minutes later: "He's loved as a person and he's hated as a person and then [people] transfer this over to his paintings."

Much like *Huey Long* before it, *Thomas Hart Benton* successfully captures the many sides of an individual who cannot be easily explained. Ken Burns's decade-long interest in quasi-biographical histories was now evidently moving in the direction of these two self-contained and more detailed character studies, leading the filmmaker to seriously consider producing a continuing series of similar biographies in the future called *American Lives*. In this historical biography, Benton as a man and an artist is vividly brought to life through his paintings, his own recorded commentaries, and a wide spectrum of opinions expressed by family members, friends, advocates, and critics of his work. Benton's sister, Mildred Small, for example, reminisces with a twinkle in her eye that "my father called him 'big I.'" Writer and acquaintance Dan James also remembers with affectionate humor that "Tom Benton was a small man with a vivacity and pugnaciousness of a bantam rooster."

Burns organized *Thomas Hart Benton* both chronologically and geographi-cally, following the protagonist's turbulent coming of age and maturation as an artist during the first 50 years of his life in the three opening chapters entitled "Neosho" (his birthplace in Missouri), "New York" ("where he worked like a whirlwind and fell in love with the only woman who could stand a lifetime with him, Rita Piacenza, who adored him, married him, and managed him"[77]), and "Kansas City" (where he came back to settle and paint his 1936 masterpiece *A Social History of Missouri*, which now adorns the interior of the state capitol at Jefferson City). The final three chapters basically chronicle Benton's decline in influence and his lifelong persistence as a painter and a chronicler of America. Chapter 4, "Years of Peril," begins with World War II and travels through the next two decades as an intemperate Benton is seen lashing out at "the limp-wristed world" of art. "Regionalism was dead," announces narrator Jason

Robards in a voice-over, "and realism out of fashion." Abstract expressionism is in instead, and, ironically, one of Benton's former students, Jackson Pollack, emerges as its most celebrated practitioner. Chapter 5, "Outliving Your Enemies," emphasizes Benton's enduring commitment to his art ("The only way an artist can personally fail, Benton said, is quit work," recounts Robards again, "Thomas Hart Benton never quit"); and, finally, a brief 6-minute cinema verité-style coda, "The Benton Bash," records the April 15, 1987 posthumous birthday party for the painter at Kelly's bar in Kansas City, where friends and admirers gather annually to celebrate the artist and his paintings.

Thomas Hart Benton specifically examines the differences of opinion that still exist among art professionals over the quality of Benton's work and his continuing legacy. Henry Adams is frequently employed throughout the film to provide the larger context: "He called himself a Regionalist, but he didn't paint one region of the country. He really roamed the whole country from New York to Hollywood, to the obscure parts of the Ozarks in the deep south, to the big open spaces of the Great Plains. Benton really got all over America." Narrator Jason Robards, too, regularly discusses the ongoing controversies in the art world about Benton as he reads detailed passages about these disputes from Geoff Ward's literate script: "Abstractionists denounced Benton for trying to tell stories with his paintings. Social Realists denounced the stories he was trying to tell, attacking him for not explicitly blaming capitalism for the Depression. Benton was unmoved. It was the sheer energy of America that compelled him."

Burns, furthermore, allows both Benton supporters and detractors many opportunities to express their pro and con views over the course of the film. A characteristic dialogue includes Lloyd Goodrich, former director of the Whitney Museum, who notes admiringly that "here's a man who took the whole face of America and tried to make a work of art out of it"; Hilton Kramer, the one-time art critic of the *New York Times,* who counters that "25 years ago one never dreamed that there would be a revival of Benton . . . his work . . . seemed safely dead"; and artist Vincent Campanella, who then presents the more populist view that "to understand Benton, you have to see America too." Hilton Kramer once again rebuts this perspective with his own high modernist evaluation that Benton "really shrank away from what I consider the largest tasks of art in this century. He had a glimpse of it in Paris and New York. He made an attempt at it and he wasn't equal to it, so he packed his bags and went back to where he could be a big figure in a small world." Henry Adams, in the end, appeals to the audience to take a deeper look underneath the surfaces of Benton's paintings to his always perceptive and ambivalent interpretations of American life and culture: "Benton began to represent the midwest in his painting, paralleling that of Mark Twain in writing, and like Twain, at first what looks simple in Benton's work ends up being complicated."

As a growing body of work, Ken Burns's first five major productions— *Brooklyn Bridge, The Shakers, The Statue of Liberty, Huey Long,* and *Thomas Hart Benton*—are well told quasi-biographical and biographical histories that consistently incorporate multiple viewpoints as a matter of course, reflecting the filmmaker's own intellectual curiosity and his liberal pluralist vision of America. Through trial and error, he "found that [the film] medium is superbly constructed to be able to comprehend many diverse perspectives."[78] *Thomas Hart Benton* went far beyond the usual sluggish pacing, solemn stylistics, and oversimplification of ideas that characterize most art films. It was warmly received during its theatrical premiere as part of the opening ceremonies for the "Thomas Hart Benton: An American Original" exhibition at the Nelson-Atkins Museum of Art in Kansas City on April 13, 1989. Its PBS debut came on November 1, 1989 during the heightened attention of a new television season. By that time, Burns had already completed *The Congress* too, marking his fifth feature length documentary in eight years since starting *The Shakers* during the late summer of 1981.

Research on *The Congress* officially commenced in September 1986 just as the production phase of *Thomas Hart Benton* was heating up, after four months of preparation.[79] As part of their initial support for The Civil War Film Project, the staff at WETA-TV had provided Burns and his colleagues with office space in Washington, D.C., during 1985, which afforded them quick and easy access to many of the major research archives and nearby battlefields for filming. Programming planners at WETA were exploring several production proposals at the time to commemorate the upcoming bicentennial of the United States Congress on March 4, 1989. Such a project both complemented WETA's national profile as one of the most important entry stations in the public television network as well as its local mission as a public service provider for the local community which, after all, included the federal government. As a result, Burns joined in these discussions as a likely participant. He suggested a project with a similar format to *Brooklyn Bridge* and the recently released *The Statue of Liberty. The Congress,* in this way, offered a topic with a rich and varied history. The film also featured the Capitol, a symbolic landmark of great national significance. The History of the Congress Film Project was formally launched by WETA on September 2, 1986.[80] Ken Burns would serve as the executive producer and director of the film, while David McCullough would be the narrator and head writer, with Bernie Weisberger also assisting in the scripting process as he had done earlier on *The Statue of Liberty.*

The overall 88-minute plot structure of *The Congress* is arranged into six chapters, five of which are quasi-biographical in nature. Like *Huey Long* and *Thomas Hart Benton* before it, this latest production by Burns was originally planned as a 60-minute documentary. Unlike those two earlier historical biographies, though, *The Congress* features an epic-sized cast spanning 200 years of American history. This film was also supposed to review all the landmark legislation and investigative actions that took place on Capitol Hill over time. Fitting this exceedingly ambitious agenda into the space of one single program was, indeed, a tall order. The highest priority of any historical documentary is to strike the right balance between delivering the needed factual information while always making sure that the drive of the narrative is strong and sustained. "The plurality of the subject made it difficult for concise storytelling," admits Burns, "the challenge was to see the whole without sacrificing stories of people and moments."[81]

All told, 26 members of the House and Senate are featured to some degree or another in *The Congress,* although other individuals are also mentioned in passing, such as James Madison, John Adams, and George Washington toward the film's beginning. The plot is smoothly constructed on the whole, but getting everything in and everybody mentioned apparently took precedence over building dramatic tension. A case in point is the second chapter, "The Debaters," which begins in 1819 and extends through the Civil War in 1865, highlighting the achievements of Henry Clay of Kentucky, John C. Calhoun of South Carolina, Thomas Hart Benton of Missouri, Sam Houston of Texas, John Quincy Adams of Massachusetts, and Daniel Webster of New Hampshire and later Massachusetts. These historic legislators deserve their rightful attention, as does the moral and political struggle that continued for decades over the question of slavery. Still all six of these figures as well as the partisan build-up to the Civil War are rapidly surveyed in only 16 minutes and 30 seconds.

Other sequences cover even more ground in shorter amounts of time. Chapter four, "The Progressives," for instance, introduces 12 legislators in 13 minutes and 20 seconds, while chapter six, "The Managers," addresses civil rights, Vietnam, and Watergate in less than 14 minutes.[82] Obviously the sweep of these chapters could easily serve as the basis for separate individual episodes if the history of Congress was ever to be redone in a longer multipart format. Five of this film's six chapters, in fact, concern one or more of the nation's legislative milestones, such as the Louisiana Purchase, a host of measures initiated by the Progressives, and the singular impact of the New Deal. *The Congress,* in general, acknowledges these seminal occurrences but mainly skims over their surfaces without exploring them in any real depth or detail. Burns's signature editing clusters, most importantly, are either minimized or absent altogether. This historical documentary, as a result, lacks the texture of the producer-director's best work, since the multiple viewpoints that usually bolster his narratives are unfortunately missing this time around.

Another goal of this project was to examine the allegorical significance of the Capitol. Burns devoted approximately half of the hour-long plot structures of both *Brooklyn Bridge* and *The Statue of Liberty* to this purpose. The much larger scope of *The Congress*, however, made such an allotment of time impractical when weighed against the filmmaker's other priorities; thus, the Capitol's review takes place within a relatively brief 7-minute 30-second chapter entitled "The Hill," although live footage of this historic building is interspersed throughout the film. Burns again takes the viewer on a high-speed 90-second pixilated tour of the physical grounds, as he did in *The Statue of Liberty*, following narrator David McCullough's offscreen explanation that "it's like a little town or a small medieval state. There's a bank, there's a post office, there's a subway. It has its own police force, and something like 44 doorkeepers, no end of clerks. There's a carpentry shop, an upholsterer, a resident architect, a resident physician. There's a daily newspaper, *The Congressional Record*. There are restaurants. It's a world within a world."

To be sure, the cinematography in *The Congress* is elegantly framed and warmly majestic, befitting the formality and beauty of the Capitol. Burns was accorded an unusual degree of freedom around the building and the surrounding premises, along with the film's other cameramen, Allen Moore and Buddy Squires. "They gave me incredible access when I made my history of the Congress," Burns recalls, "but it was as if I were one of them, that they felt that I knew enough about the dynamic that I was an initiate."[83] In getting up to speed on the topic, Burns wisely scheduled an extended production period because of his concurrent and all-consuming commitment to *The Civil War*. To keep that much longer and more ambitious miniseries on schedule, "[he] rented an apartment on Manhattan's Upper West Side for the work week [during 1988 and 1989], returning to his family in New Hampshire only on weekends. Working fifteen-hour days, he soon found that he needed to abstain from alcohol and incorporate regular exercise into his routine to provide himself with the necessary stamina. He [even] gave up his social life entirely."[84] There were times in that period when Burns understandably felt overcommitted; his characteristic response under such pressures was simply to work even harder.

The Congress debuted before an enthusiastic audience of House and Senate members, their families, and guests at the National Theatre in Washington, D.C., on March 13, 1989. Its initial telecast on the public television network was also a popular success a week later on March 20, joining Burns's other television histories as among the most-watched nonfiction films on PBS during the 1980s. History had been largely absent from television only a few years earlier. Most network executives in both the private or public sectors had believed that historical documentary films and mini-series were poorly suited to TV as a medium and as an industry that required millions of viewers as its main indicator of success. Yet Burns's work was one of the few exceptions to this programming rule. He had created a special place for himself in the independent filmmaking world and the Public Broadcasting Service through the quality of his historical documentaries, his relentless work ethic, and the sheer force of his will to excel.

At first it may appear that Ken Burns embraced a wide assortment of subjects in his first six television histories—a bridge, a nineteenth-century religious sect, a statue, a demagogue, a painter, and the Congress—but there are a handful of underlying common denominators that bind this medley of Americana together. First of all, Burns's pluralist documentaries are fundamentally liberal on social issues, such as civil rights, while at the same time traditional when it comes to core American values and the nation's institutions. Second, his subjects tend to be majoritarian in outlook, reflecting his sympathy for those representative themes, issues, and ideals that somehow bring the country together rather than divide and segregate from a special interest point of view. Third, Burns is equally interested in history and myth. He, therefore, creates a series of documentary films as morality tales, drawing upon epic events, landmarks, and institutions of historical significance, populated by heroes and villains who allegorically personify certain virtues and vices in the national character. Fourth, he increasingly incorporates the American original characterization into his nonfictional storylines, which reveals his abiding faith in the ability of individuals to initiate historical change and assert a profound impact on the national culture. Fifth and, finally, Burns produces his television histories for a broad and diverse popular audience. In his own words, "I like the challenge of speaking to the whole country. I like the idea that there may be just a few things in our ever diverging country that we can still share together."[85]

Looking back at this juncture of his career, Burns further reflects,

I couldn't imagine things could get better for me. *Brooklyn Bridge* had been nominated for an Oscar. My next project on *The Shakers* was extremely well received as was my film on *Huey Long* which enjoyed a theatrical release. *The Statue of Liberty* was also nominated for an Oscar. So on the eve of the release of *The Civil War* series, in the fall of 1990, I felt that I was at the very top of what I could do. And then *The Civil War* hit and it was almost as if I was born yesterday.[86]

Even though Burns's prospects were rising even higher now as a result of his new-found level of popular success, he spent little free time resting on his laurels. Two of the few things that remained constant for him over the next couple of years were his love of American history and his undiminished passion for producing and directing films.

The Creative Team as Historian: Inside the Production Process on *Empire of the Air: The Men Who Made Radio* (1992)

LAUNCHING THE RADIO PIONEERS PROJECT

Empire of the Air was a gift. Usually all of the films that I've done have sort of come out of me, but in a couple of instances, someone else has suggested them and that was the case here.

—*Ken Burns, 1996*[1]

Empire of the Air: The Men Who Made Radio was Ken Burns's eighth major PBS special, premiering on January 29, 1992. Its production history offers a glimpse into Burns's working methods at Florentine Films during an extraordinarily heady period for both the producer-director and his still-struggling company. This historical documentary was conceived and created while *The Civil War* was being edited and later released to wide attention and acclaim, a period when Burns's professional profile changed dramatically as he became a national celebrity virtually overnight. The heightened work environment and reaction surrounding *The Civil War* also affected *Empire of the Air*'s shooting and assembly stages, in minor ways at first, but then more tangibly after the summer of 1990. The production of this PBS special, in retrospect, provides a revealing object lesson into the shared authorship which typically occurs when creating mediated history on film for television.

This documentary as made-for-TV history originated with Tom Lewis, a good friend and sometimes collaborator of Burns, who briefly served as head researcher on *The Civil War* before dropping out to pursue his rapidly growing interest in Edwin Howard Armstrong and the early history of radio broadcasting. Lewis first became aware of Armstrong quite casually, when his wife showed him a *New York Times* article on the inventor. The story fascinated him enough to research and write a preliminary article, "Radio Revolutionary: Edwin Howard Armstrong's Invention of FM Radio," which David McCullough recommended to *American Heritage of Invention & Technology*, where it appeared in the fall 1985 issue.[2] That publication brought Lewis to the attention of several close associates of Armstrong, including his patent lawyer Dana Raymond, "open[ing] up a lot of doors to people I had never heard of, " Lewis recalls. "I played with the idea for a long time, and then I figured out it wasn't just a story about Armstrong, but it was a story about Armstrong and de Forest and Sarnoff—and it was three people. And then I felt I had a structure I could work with," as he decided to develop the historical narrative into both a book and a companion film.[3]

Ken Burns was initially unpersuaded by Lewis's proposal of producing a documentary version of what at the time was called "Radio Pioneers," being fully preoccupied with the enormous challenge of seeing *The Civil War* through to completion. Then on Labor Day 1987, while visiting Burns's home in Walpole, New Hampshire, Tom Lewis once again outlined the whole story, and together they literally "drew a picture of how the film would work." Lewis remembers, "we structured the whole thing on a piece of 8½-by-11 inch paper turned sideways with the three movements of the film."[4] Interestingly, those three sequences—"The Men and the Idea" (introducing Lee de Forest, E. Howard Armstrong, David Sarnoff, and the notion of radio broadcasting), "Broadcasting" (continuing the interaction of the three main characters along with the extraordinary rise and popular success of radio), and "Tragedy and Triumph" (depicting the three very different fates of Armstrong, de Forest, and Sarnoff against the much larger backdrop of American business, technology, and society)—remained basically intact throughout the entire production process with only slight modification.

Burns and Lewis, most importantly, recognized the inherent drama in E. Howard Armstong's life story, much as they had responded years earlier to the histories of John and Washington Roebling building the Brooklyn Bridge and

Mother Ann Lee inspiring her followers in the Shakers. "What we're talking about is a great American story," Lewis explains, "a story about friendship, trust, and ultimately betrayal. It had Lee de Forest who was about as good an inventor as Gyro Gearloose. And then it had an authentic genius in Armstrong. And it had the man [Sarnoff] who took those inventions and engineered the sales of those devices to make a lot of money and that to me was an exciting story."[5] Burns, likewise, reminisced during a February 1993 interview: "A very dear friend of mine, Tom Lewis, was writing a book about these three men as a wonderful backstage drama. And he wanted to make a film as well and he enlisted my aid. And as I was helping, I became interested in the story. I realized that this was something that I wanted to do, too, and he was more than happy to have me take over his film project."[6]

Empire of the Air took more than five years to prepare, fund, shoot, assemble, and distribute, beginning with that joint agreement between Burns and Lewis on September 7, 1987 and running through its distribution on video and its rebroadcast on public television on March 31, 1993. Burns was deeply immersed in *The Civil War* when the Radio Pioneers Project was first launched, and he was similarly absorbed in *Baseball* (PBS, 1994) when *Empire of the Air* was finally released to syndication. Burns was still the major force on all of his historical documentaries, but his company had matured to a point where he was now working on multiple projects simultaneously, collaborating more than ever with an array of coproducers who both shared his interest in history and adopted his well-known and well-established signature style. *Empire of the Air*, additionally, appeared first as a book, then as a major public television special, and, lastly, as a radio play, providing a representative example of how history (like any other narrative genre) is regularly adapted across multiple media forms these days and, in turn, is directed toward varying segments of the American mass audience.

CLIMBING THE MOUNTAIN OF MONEY

As the book evolved, so the film progressed along with it.
—*Tom Lewis, coproducer, 1993*[7]

The Radio Pioneers Film Project began in earnest on Labor Day 1987, the day Ken Burns decided to commit himself fully to the production along with Tom Lewis. As is the case with all independent production companies, the first line of business is cobbling together the needed funding, usually from a variety of sources. Up to this point, Burns's track record was accomplished, having completed *Brooklyn Bridge, The Shakers, The Statue of Liberty,* and *Huey Long,* but raising money remained a long and protracted undertaking for Florentine Films, especially in the days before the popular success of *The Civil War.* Burns first approached Ward Chamberlin, then President of WETA-TV 26 in Washington, D.C., which was also sponsoring two of Burns's other productions at the time, *The Congress* and *The Civil War.* The reaction from Chamberlin and his colleagues at WETA to the "Radio Pioneers" proposal was enthusiastic and supportive. The station's senior staff appropriated $20,000 in start-up monies, directed its development office to begin soliciting corporate underwriters, and, most importantly, agreed to be the entry station for the national telecast. Lewis then secured another $5,000 grant from Mitchell Hastings, president of the Armstrong Memorial Research Foundation at Columbia University, to further defray preproduction costs.

Preparation proceeded on three additional fronts simultaneously once seed money was in place. First, Tom Lewis continued working on his book, essentially establishing the research and narrative agenda for "Radio Pioneers" in both its evolving written and film forms. He visited all the key archives associated with Armstrong, de Forest, and Sarnoff, finding a wealth of primary source materials in print, while at the same time scouting for possible photographs, stock footage, and sound recordings for the upcoming film.[8]

Second, Burns and Lewis assembled their production team throughout 1988. Several members of Burns's stock company at Florentine were enlisted straightaway, including Geoffrey Ward as head writer, Buddy Squires as cinematographer, and Paul Barnes as editor. Ken Burns would be the executive producer, and he and Lewis recruited Kenneth Fink to line-produce and direct. Fink had recently created two award-winning documentaries, *Between Rock and a Hard Place* (1981) and *The Work I've Done* (1984), and he was currently employed as a segment producer on CBS's *West 57th Street* (1985-1989), a prime-time news program aimed at audiences under 40. Garrison Keillor, who was then acting as one of the offscreen voices on *The Congress* and *The Civil War,* had preliminary talks with Burns about narrating; and Tom Lewis was appointed coproducer with Burns and Fink, providing introductory background materials to scriptwriter Geoff Ward at this juncture and coordinating efforts with WETA and Florentine Films on fundraising.

The project director for WETA was Tamara Robinson, then vice president for cultural affairs at the station. Burns and Lewis had known Robinson since the late 1970s when she had been an early advocate of *Brooklyn Bridge,* helping to nurture that production at Thirteen/WNET's Television Laboratory. Hers was the delicate job of shepherding "Radio Pioneers" through PBS's institutional apparatus, promoting its financial support with the Corporation for Public Broadcasting (CPB) and procuring the eventual carriage agreement for a core telecast on the public television network. A funding request for $125,000 was submitted to CPB.

Third, Lewis and Burns began lining up a working group of scholars, writers, and witnesses who would advise the project and provide commentary on film. All of Burns's documentaries are modeled on the existing historical record and designed to bridge public interest in the subjects he chooses with the combined findings of the academic and literary communities. "Radio Pioneers" is no different in this regard, relying mostly on five principal consultants: Erik Barnouw, the preeminent historian of broadcasting; Kenneth Bilby, former executive vice president of RCA and close associate and biographer of Sarnoff; Susan Douglas, then associate professor of American studies at Hampshire College and a social historian of radio; Bruce Kelley, licensed radio amateur and former curator of the Antique Wireless Association (AWA)'s Museum; and Bruce Roloson, then President of the Antique Wireless Association, one-time engineer, and expert on de Forest and his development of the vacuum tube.[9] Twenty other individuals were also contacted—a diverse and complementary group of radio professionals and personal acquaintances of Armstrong, de Forest, and Sarnoff agreed to be interviewed on camera.[10]

Now armed with a more thoroughly researched understanding of their subject, a sound and viable narrative structure, an experienced production team, and a cadre of scholarly and professional advisors, Burns and Lewis collaborated by telephone and overnight mail during the late winter of 1989 on a $477,334 grant application to the Public Programs Division of the National Endowment for the Humanities (NEH).[11] Lewis did most of the actual writing of the proposal, since Burns was largely engaged in editing *The Civil War.* Tamara Robinson also reviewed a draft of the working document since it was being submitted under the auspices of WETA. On Tuesday, March 14, the "Radio Pioneers" team received word that CPB had awarded them an initial $75,000, just as Tom Lewis was literally wrapping up the final version of the proposal. Lewis quickly added an accompanying letter to the NEH division head explaining that Florentine Films had now raised $100,000 of the now projected $627,334 budget, thus underscoring the wider sponsorship now materializing for the project.[12] A copy of *Huey Long* was also submitted as a representative example of past work for the NEH review committee. Attention then shifted to other matters as Ken Fink withdrew to accept a series of directing assignments on prime-time TV, since fundraising was proving more drawn out and uncertain than the more immediate employment opportunities available in the commercial sector.[13]

Morgan Wesson, a film curator at the Eastman House in Rochester, New York, soon filled the breach. Wesson graduated from Hampshire College in 1974, a year ahead of Burns, and then attended MIT for advanced work in film, as did several other of Jerome Liebling's students during the mid-to-late 1970s and early 1980s. He and Tom Lewis were also members of the Antique Wireless Association, which is dedicated to the study and preservation of old radio equipment and research materials and operates a museum in Holcomb, New York. At one of the annual AWA conventions, Lewis told Wesson the story of E. Howard Armstrong over dinner. Wesson was so captivated by the tale that he contacted Burns "and asked if [he] could please come along." Burns was a little wary at first, "you know someone you [go] to school with, an old buddy," but Wesson's background as a media archivist and historian ended up complementing Burns and Lewis so well that he eventually became a coproducer on "Radio Pioneers." As Wesson recounts, "they were both busy men. Ken had just ended editing *The Civil War* and he was starting to market it, and Tom has a full-time teaching position, so I became their eyes and ears—the hunter-gatherer—and went out and got the audio and film and then brought it back for the look-see."[14]

Lewis brought Wesson up to speed on all the main repositories and traveled with him to the Sarnoff Research Center in Princeton, New Jersey, where they scoured the collection for promising audio and visual clips that could be integrated into the forthcoming film. "Morgan was going out into all these old radio collections and gathering sound elements," Lewis recalls, "and then all of the film elements. He got almost all of them. He's the one who did 99 percent of that work and did a great job."[15] Wesson, too, was fascinated with the mostly hidden history surrounding early radio and was stimulated by the chase of uncovering the actual account: "The story of Howard Armstrong is just one of those undiscovered jewels of technological history. Every time people turn on their radio, it would be nice if they remembered him."[16] The "Radio Pioneers" creative team was well prepared and ready to start production by the summer of 1989; their funding needs were eased considerably in one fell swoop when Tamara Robinson of WETA was notified in August that the NEH had awarded them the entire $477, 334. Now the personnel at Florentine Films could devote most of their attention to constructing the historical narrative out of what Ken Burns characterized as "a very dark twentieth-century story."[17]

LEFT-HAND, RIGHT-HAND PROCESS

Ken and Paul are a dynamic combination. That is just a remarkable blending of talents and sensibilities. These two men see eye to eye yet they challenge each other constantly and they have the same feel for what makes a narrative out of history. There's nobody out there who does this better than Paul Barnes in the editing room and Ken out there forming and shaping and putting the production together and seeing it right through. He is just relentless in his vision.

—Morgan Wesson, coproducer, 1999[18]

Burns now assumed the role of producer-director while Wesson continued searching for additional aural and visual raw materials that could be considered for the upcoming film, and Lewis intensified his interactions with Ward, who was fully engaged in scripting "Radio Pioneers": "I fed him my book and Geoff turned it into a script; and I would supply lines and graphs and things like that, but it was Geoff who really did the scriptwriting on it. I was handing him chapters as I was writing them. It worked out very well actually."[19] With cameraman Buddy Squires and the two coproducers, Burns was concurrently conducting the 25 filmed interviews, which were quickly transcribed and given back to him. Burns would then review this material carefully, adding detailed directions on what interviewees and quotes looked most promising. On veteran radio dramatist Norman Corwin's transcription, for example, Burns wrote, "Geoff—This is our Shelby, use it all—KB" (referring to how the popular historian and gifted storyteller Shelby Foote was utilized throughout *The Civil War*).[20] These transcriptions were, in turn, mailed back to Ward for his reference and integration into the script.

Ken Burns refers to this phase of production as his "left-hand, right-hand process" of independently shooting and advancing the script, "leaving things open to discovery as long as possible."[21] All of these individual lines of input are later assembled by Paul Barnes (and his chief assistant on "Radio Pioneers," Yaffa Lerea) and painstakingly fine-tuned over many months (and occasionally years in the cases of *The Civil War, Baseball,* and *Jazz*) of collaborative discussions in the editing room, which is where much of the critical decision making occurs on all of Burns's historical documentaries. Ultimately, though, "Ken was always the final arbiter," remembers Morgan Wesson.[22]

Preliminary editing and script development slowed somewhat because of the publicity efforts and extraordinary response to *The Civil War* and all the resulting public demands. Burns completed an arduous two-month promotional tour on Sunday, September 23, 1990, the day of the series premiere, and then in Tom Lewis's words, "*The Civil War* broke like a tidal wave and really took Ken away from the production of *Empire,* which was fine because he had a good backup team. He had Paul and me and Geoff and he had Morgan."[23] Wesson similarly recalls, "After *The Civil War* it was comic-book stuff. Ken was on the road a lot in those months. He was out lecturing to put money back into his struggling company [Florentine Films]."[24] "Even when Ken was knee-deep in *The Civil War,*" Lewis adds, "he would still say do this, do that, do this, and that's one of the powers of Ken that he really has that ability."[25]

The whole assembly process gained renewed momentum when Geoff Ward completed an initial installment of his "Radio Pioneers" script on December 13, 1990.[26] The fundamental composition of this 53-page first draft (including the narration, offscreen readings, and onscreen commentaries) conformed to the three-act structure devised by Burns and Lewis on Labor Day 1987. The basic story features were similarly adapted straight out of Lewis's developing book manuscript.[27] "Radio Pioneers" focuses primarily on three men: Lee de Forest, the self-proclaimed "father of radio" and "grandfather of television," whose 1906-1907 discovery of the audion or three-element vacuum tube made radio-wave detection, rudimentary amplification, and reception possible, thus auguring the birth of modern electronics; E. Howard Armstrong, whose three greatest inventions—the regenerative circuit in 1912, the superheterodyne circuit in 1918, and a complete FM system in 1933—serve as the technological foundation for radio broadcasting as we know it today;[28] and David Sarnoff, the Russian immigrant who became one of the most powerful industrialists in America, dedicating his fifty-year career at the Radio Corporation of America (RCA) to the commercial realization and widespread acceptance of both radio and television.

Hayden White has theorized that the work of traditional historians is never a matter of just letting a fully recoverable past speak for itself, but a far more proactive process of "emplotment of historical information" into the form of recognizable "narrative and literary tropes."[29] The shorthand of screen language as history employs similar fictional conventions, often in a more explicit fashion than written history. In Ken Burns's own words, "*Empire of the Air* is a twentieth-century story which borrows themes from Charles Dickens and Thomas Hardy and includes all the ingredients of Greek tragedy—pride, greed, deceit, and even despair and death."[30] The plot line of "Radio Pioneers" chronicles the history of radio broadcasting as a technology and social phenomenon from a mostly biographical perspective. The documentary focuses on the three men whom the filmmakers identify as most responsible for radio's invention and development. The scope of their story begins roughly in 1906 with de Forest's discovery of the audion and concludes in 1954 with Armstrong's suicide, although pre-1906 and post-1954 details are also provided as a way of fleshing out the main characters' boyhoods in the beginning and of tying together all the loose narrative ends during the film's final five minutes.

The three most important assumptions that Burns and his creative team at Florentine Films advance in constructing the "Radio Pioneers" plot are that, first, Guglielmo Marconi invented wireless telegraphy and not radio as most people in the general population probably assume.[31] Second, if anyone deserves to be remembered as the "father of radio," it is E. Howard Armstrong, since his multiple contributions to the creation of radio broadcasting (see note 28) are far more enduring and fundamental than Lee de Forest's invention of the three-element vacuum tube, a technology which he never fully understood at the time that he made the discovery. And third, the years in which the "Radio Pioneers" story happens mark an essential shift away from the nineteenth-century concept of the lone American inventor to the full adoption and ascendancy of the corporate-sponsored approach to research and development as embodied by David Sarnoff at his RCA Lab in Princeton, New Jersey. As Tom Lewis describes in his book: "No longer would an individual—a Thomas Edison or a Charles Goodyear—work alone or with a few assistants to make great discoveries. Now groups of anonymous technicians would labor for giant corporations."[32] Ken Burns likewise asserts that "because [these three men] lived in the decades [when] corporations became more and more sponsors of inventions, lone inventors got submerged in the corporate reality, and they in turn became more ruthless, more paranoid, more self-centered—corrupted in a spiritual way. Each of our three heroes internalizes this corruption. What should be a story of genius and determination becomes a story of insecurity and ultimately tragedy."[33]

The narrative emphases of Burns and his production colleagues are readily apparent early on when they introduce the three main characters toward the end of the short seven-minute prologue, six minutes into the film. A montage of well-known celebrated announcements (e.g., "The Japanese have attacked Pearl Harbor!" "The Giants win the pennant!" etc.) plays over a close-up of an old-time radio dial filmed in a warm orange hue. The narration (provided by Jason Robards, who replaced Garrison Keillor so that the latter could be featured instead as an onscreen commentator) gradually fades in: "The world these three men helped to make was altogether new" (a form cut begins as the shot of the circular dial dissolves into another shot of a similarly shaped audion tube held by Lee de Forest), "but they were driven to create it by ancient qualities" (the camera slowly pans left within the same photograph away from the audion toward the face of de Forest) "idealism and imagination, greed and envy" (the camera stops on a close-up of de Forest), "ambition and determination" (cut to a head and shoulders shot of David Sarnoff with a resolute expression), "and genius" (cut to a close-up of E. Howard Armstrong). Then another brief radio montage fades in and ends this highly concise and effective preface with the voice of Orson Welles as Lamont Cranston delivering the famous tag line from *The Shadow:* "Who knows what evil lurks in the hearts of men?"

All of Ken Burns's documentaries typically employ a biographical or quasi-biographical approach as a more personally engaging way of animating larger historical themes and issues. The "Radio Pioneers" documentary is yet another one of Burns's morality tales, inhabited by three distinctly strong individuals who broadly embody certain strengths and weaknesses in the American character. The linkage of word and image in this short scene from the prologue subtly and succinctly suggests to the audience how they might want to think about each of the three main personalities in the film. The editorial choices, in essence, frame Armstrong as the brilliant if tragic hero, de Forest as the jealous and somewhat buffoonish pretender, and Sarnoff as the aspiring and calculating corporate man.

Act 1, "The Men and the Idea," rhetorically reinforces the first impressions created in the prologue of the protagonist, E. Howard Armstrong, and his two enduring antagonists, Lee de Forest and David Sarnoff. The section on de Forest begins with his guest appearance on *This Is Your Life* (NBC, 1952-1961), in which a fawning Ralph Edwards holds up a replica of the first audion and calls it "the miracle seed from which sprang the entire mighty structure of radio and television, sonar, radar, talking pictures, guided missiles, automation, the electronic brain computer, and long distance telephone communication." De Forest, who indeed reveled in self-promotion and hyperbole all of his life, is framed from the outset within the smoke and mirrors of this prime-time testimonial, thus imbuing the short biographical sketch that follows with an air of skepticism and ironic amusement.

David Sarnoff's rags-to-riches story at the end of Act 1 is likewise introduced with interpretive flair. The scene starts with Sarnoff being serenaded by Frank Sinatra to the tune of "The Lady is a Tramp" at his fiftieth wedding anniversary banquet on July 4, 1967: "Telegraphy showed the General the way, Took infant radio and taught it to pay, And he began what is now RCA, That's why the gentleman is a champ, To radio what Edison was to lamp, That's why the gentleman is a champ!" Then the scene cuts to newsreel footage of Governor Nelson Rockefeller eulogizing Sarnoff in 1971: "Who was this man? What was he? Others looked at radio and saw a gadget. David Sarnoff looked at radio and saw a household possession capable of enriching the lives of millions. David Sarnoff was so thoroughly American; no fiction celebrating the American dream could begin to approach his life." Sarnoff is instantly identified with the rich and powerful, thus deflating the ensuing account of his remarkable climb out of abject poverty. As his biographer Kenneth Bilby summarizes onscreen: "No one has ever started from a lower pane of life and gone further, in my judgment, than Sarnoff did. He reached this country when he was nine years old, an immigrant boy who didn't speak a word of English, but in 30 years he was president of one of the most powerful companies in America—RCA."

Armstrong's early history is also briefly outlined in Act 1 between the sections on de Forest and Sarnoff, spanning his boyhood in Yonkers, New York, his 1912 discovery of the regenerative or feedback circuit, and his 1914 meeting with the then-chief inspector of the Marconi Wireless Company, David Sarnoff. This seven-minute scene starts with four short onscreen commentaries that, once again, underscore how the "Radio Pioneers" creative team stands on Armstrong, beginning with Jeanne Hammond: "It makes me very sad that his name isn't better known. I remember as a child talking about him and saying, 'My uncle invented radio'"; engineer Frank Gunther: "I believe that when history is written, all the history on this, Howard Armstrong should be remembered as the greatest inventor of the twentieth-century"; engineer Robert Morris: "He was the best development engineer I ever knew"; and engineer Loren Jones: "He was an extraordinary person. He exuded and carried a sort of aura of invention. He was a true genius and there aren't very many in the world today." The stage is now set for the balance of the historical narrative.

Tom Lewis describes the classical pedigree of the plot line in the prologue of his book: "Each of these men acted as a protagonist in a drama with Olympian overtones, replete with the elements of classic tragedy: anger and distrust; hubris and blindness; destruction and death."[34] This epic conflict in Act 1, "The Men and the Idea," is produced mostly over the patents struggle between Armstrong and de Forest concerning the regenerative circuit. The tension is generated less by details surrounding the court proceedings, however, than by the stark contrast being drawn between two very different personalities: media historian Susan Douglas aptly describes Lee de Forest as "a wonderful character, a churl, a roué, a cad, a thief, a romantic, and yet, probably the man most responsible for bringing radio broadcasting to the American public." Armstrong, on the other hand, is viewed as tall, serious, intensely self-contained, and a genuine electronics prodigy; he is also described by narrator Jason Robards as "aloof, sarcastic, altogether too willing to question basic assumptions in front of his fellow students [at Columbia University]." Neither man has any inclination of compromising with the other, as Armstrong wins the first few lower court rulings in Act 1. Robards continues: "For the time being, everything seemed to be going Armstrong's way. Sale to RCA of his brand-new discovery, the superregenerative circuit, had made the inventor the single largest stockholder in the company. And he was eagerly courting David Sarnoff's secretary, Marion MacInnis." The act then concludes with Armstrong marrying MacInnis in 1923 while a lively and upbeat version of "Love Her By Radio" plays on the soundtrack, temporarily punctuating the story's mood.

Act 2, "Broadcasting," continues the unfolding relations among Armstrong, de Forest, and Sarnoff, but shifts more attention to the "programs [from] the 'golden age of radio' [and] the various entertainers from the thirties and forties who captured the imagination of millions of listeners." As Lewis and Burns proposed in their NEH grant proposal, "this documentary . . . focuses primarily on those who made radio possible and the complex effect the new medium had upon the culture as a whole."[35] All of Burns's work strives to suggest how individual stories and anecdotes from history can be interpreted as signs of the larger American experience. In the case of *Empire of the Air* specifically, the filmmakers struggled for months to assemble direct linkages between the private lives of their featured characters and the broader currents of radio's more public culture with mixed results.

Two major reasons surface as to why this strategy didn't quite work. First, none of the three radio pioneers was personally involved in the electronic consumer culture created by broadcasting. The narrator reports that "the three men most responsible for radio's popularity rarely listened to it. David Sarnoff was too busy building his broadcasting empire. Lee de Forest could not endure the commercials. Howard Armstrong didn't listen to radio because he was so annoyed by the static." Their complex relationships with each other and their habitual legal wrangling were, ironically, far removed from the golden age of radio programming as understood and experienced by the vast majority of ordinary American listeners.

Second, this apparent divide between the radio pioneers and radio culture is reinforced in the full commentaries provided by Norman Corwin, Garrison Keillor, and Red Barber (as well as other broadcast personalities and professionals, such as Eric Sevareid and Robert Saudek, who never made it into the final print). Each of these interviewees express little or no knowledge of Armstrong in particular, and only a general awareness of de Forest and Sarnoff. As a result, they were unable to directly link the biographies of the three men with any broader societal dynamics related to radio. Consider, for example, a representative exchange between Burns and Garrison Keillor from the outtakes of his filmed interview:

KB: Why are the men most responsible for radio unknown?

GK: You're asking me about something I really don't know.

KB: Why doesn't Garrison Keillor know who invented radio?

GK: I guess I was happy to believe the legends of, you know, Marconi that sort of emanated from somewhere in the nineteenth-century."[36]

That Ken Burns recognized this emerging bifurcation in the "Radio Pioneers" narrative is confirmed by his written responses to the first two drafts of the script in December 1990 and January 1991, as well as by his main recommendations after viewing Paul Barnes's initial rough cut of the film in January 1991: "We need to set up earlier good friendship of A [Armstrong] & S [Sarnoff]—we lose deF [de Forest] altogether" and "Keep alive storyline during Broadcasting."[37] Act 2, in turn, was dramatically streamlined by Barnes in the editing room during the spring months as more than half the commentaries, sound montages, stills, and live footage referring to both the golden age of radio and the coming of television were eliminated in a finer cut of the film and a corresponding fifth draft of the script in June 1991. Thirty total hours of footage were eventually winnowed down to three by June 1991, as the "Radio Pioneers" creative team deliberated among themselves over what would be the most appropriate length and format for the final release.

In the release print, Lee de Forest does make a brief appearance toward the beginning of Act 2, when in one of those Dickensian twists of fate that Burns and Lewis had originally envisioned, he surprisingly wins the patents suit over the regenerative circuit in the Supreme Court "on an arcane question of legal language." Armstrong's friendly and mentorlike professional relationship with the slightly older but much more worldly Sarnoff is also more sharply defined, leading Erik Barnouw to recount in the first few minutes of Act 3, now retitled "A Little Black Box," that "Sarnoff had once said to him, 'Why don't you come up with a little black box to eliminate static?' He had already become a millionaire from patents he had sold to RCA, and he got busy in the basement of Philosophy Hall [at Columbia University] to work on this thing."

Act 3 is mostly consumed with Armstrong's remarkable 1933 invention of FM radio as well as the protracted and bitter legal battle he waged with Sarnoff over its ultimate use and dissemination. Armstrong wanted FM to replace AM, a revolutionary gesture which at the time meant the wholesale replacement of the industry's entire technical infrastructure along with all home receivers totaling in the tens of millions. Sarnoff was much more committed to promoting television and thus used his considerable power at RCA to thwart the inventor at every turn. Armstrong died tragically by his own hands, jumping from the thirteenth floor of his Manhattan apartment on Sunday, January 31, 1954, a nearly broke and beaten man. Morgan Wesson remembers how difficult it was to reconstruct that climactic scene at the time: "Ken absolutely couldn't identify with what Armstrong did. He saw it as a crime against nature. I think because of that we spent a little less time there but it was beautifully done."[38] The "Radio Pioneers" narrative, in this regard, is uncharacteristically downbeat for Burns. His documentaries usually celebrate individuals of national significance who persist against great odds; Armstrong, instead, is an anomaly. Even the protagonists who die in Burns's other documentaries, such as Huey Long and Abraham Lincoln, are killed at the height of their powers by other men, thus enhancing their respective reputations as figures of historical importance and influence.

E. Howard Armstrong's legacy, in contrast, needed serious resuscitation by the time Tom Lewis and Ken Burns happened upon his life story. He was largely forgotten by most people other than media historians and electronics engineers. Armstrong was absent, for example, from two major exhibits sponsored by the Smithsonian Institution at its National Portrait Gallery ("On the Air: Pioneers of American Broadcasting," October 1988–January 1989) and the National Museum of American History ("American Television: From the Fair to the Family, 1939-1989," April 1989–April 1990). The National Portrait Gallery presentation was even cosponsored by the Museum of Broadcasting (now the Museum of Television & Radio), featuring five radio "Inventors" and "Entrepreneurs"—i.e., Guglielmo Marconi, "The real genius of the 'Wizard of Wireless'"; Lee de Forest, "The Father of Radio"; Herbert Hoover; David Sarnoff, "The Prophet of Radio and the Father of Television"; and CBS founder and guiding force, William S. Paley.[39] Granted Marconi, de Forest, and Sarnoff were talented self-publicists as well as accomplished innovators; still, Armstrong's suicide remained a source of great pain in the professional community, prompting many to just avoid thinking about his tragic demise altogether. More than three decades had passed by the time the "Radio Pioneers" film project started production. Despite whatever ambivalence Burns might have felt, his expressed intention was to revise the existing historical record and reclaim Howard Armstrong's rightful place in the history of radio.

SETTING THE RECORD STRAIGHT

Armstrong's name should be as well known as Edison's. The reason it isn't is the reason we did this film.

—*Ken Burns, 1992*[40]

Revisionism is an unusual historiographic stance for Ken Burns and his colleagues at Florentine Films. They usually approach each television special as an exploration into the historical topic at hand, holding far fewer preconceptions than they did with *Empire of the Air* in terms of how to structure the plot line and reclaim the forgotten status of their main character. Burns is a highly accomplished filmmaker, but as he often claims himself, "an amateur historian."[41] His work habits, however, do complement and reflect many customary academic practices. The relationship between Burns's creative team and the professional consultants on the "Radio Pioneers" film project was constructive and complementary. Tom Lewis first visited media historian Erik Barnouw in Philadelphia at the University of Pennsylvania, where he was midway through his extensive duties as editor-in-chief of the then-forthcoming four-volume *International Encyclopedia of Communications*.[42] Barnouw accepted Lewis's invitation to advise the "Radio Pioneers" project, being a longtime supporter of made-for-television histories and having met Burns in 1982 when *Brooklyn Bridge* won the first Erik Barnouw Prize from the OAH for Outstanding Documentary Film in American history. Barnouw's earliest contribution to *Empire of the Air* was a five-page outline of salient points and suggestions to guide the filmmakers as they immersed themselves ever deeper into research and preparation.[43]

Narrative is by far the oldest and most consistently popular way of framing history. In agreeing to produce a triple biography in three acts, Burns and Lewis were selecting a strategy of representing the past on film that was principally literary and dramatic in outlook. Burns is not as concerned with matters of historiography as are most professional historians. Still, he is well aware that there are various approaches to constructing history, and he and his colleagues take great care to be as accurate as possible in assembling their documentary stories. Above all else, though, Burns's highest priority is creating and meticulously refining a sharply defined and compelling historical narrative. In that way, too, Tom Lewis immediately recognized the rich dramatic potential inherent in the "Radio Pioneers" plot line.

> I did not go into this to be a crusader for Armstrong, though the book could easily be read that way, because in point of fact it was a book that told about a man that nobody knew about. The reason I wanted to tell the story was that it was a great story. It was a story about a great inventor and it was a story about friendship and betrayal. What we're talking about is a great American story. This is a story about America. So those were the reasons that I wrote it. It was a good yarn.[44]

Lewis and Burns set three main priorities in motion when they jointly established the narrative parameters of *Empire of the Air*. First, they choose to make a biography, which is characteristic of Burns's work in general as well as most television histories. History on TV, almost without exception, gravitates toward the portrayal of individuals from the past as tangible embodiments of much larger social, cultural, and historical concerns and trends within as compelling a plot structure as possible. Second, they, more specifically, decided upon a triple biography, which presented a far more intricate storyline featuring an ardently sympathetic lead, Howard Armstrong, who is engaged in almost continual dramatic conflict with either Lee de Forest or David Sarnoff, two deeply flawed but formidable adversaries. The "Radio Pioneers" creative team, as a result, was far more reluctant to spend considerable time on other important figures in the history of early radio, such as Marconi, Fessenden (the Canadian inventor of wireless telephony), and Paley, given the greater complexity of this three-sided biographical narrative. And, third, they were telling an American story which again shifted attention away from inventors like Hertz, Marconi, Fessenden, and other relevant figures, thus framing a longstanding international project that had involved the transnational scientific community (as well as scores of resourceful amateurs from a variety of western nations) into a largely domestic accounting of the men who made radio.

Empire of the Air was, additionally, designed to "step back from the drama of these three lives to consider just what the creation of radio meant to the culture."[45] Burns and his creative team consistently strove to emphasize the pivotal events of the de Forest, Armstrong, and Sarnoff life stories as not only turning points in their own individual fortunes but also telling indicators of much broader historical currents and issues. Erik Barnouw, for instance, frames the personal drama of de Forest's defeat of Armstrong over the regenerative circuit in the Supreme Court within a much larger framework through his commentary toward the end of Act 2:

> The struggle between these individuals becomes the struggle between corporations that have acquired their patents. The struggle gets larger all the time. And then it also becomes a struggle between nations because these inventions, while they start out being toys and hobby things, soon become important corporate assets, and pretty soon they become assets of military industrial complexes. And these small things also get swallowed up not only in corporate affairs but in national affairs.

Contextual matters emerged early on in the research, the grant applications, the initial scripting and editing, and throughout the 25 filmed interviews, particularly with the five consultants. Susan Douglas's conception of technology in *Inventing American Broadcasting, 1899-1922* is highlighted as a guiding principle for the "Radio Pioneers" team in the grant application for example: "Technology is as much a process as a thing; it is an evolving relationship between people and their environment."[46] She was encouraged, moreover, to speak extensively about consumer culture and the desires of the mass audience in her film interview.[47] As Douglas remembers,

> I have to say I have never been interviewed by anybody who was as good an interviewer as Ken Burns. Some of it is the way this team works. You know that you are having superb lighting and you can tell by the care that goes into the whole setting up of a shot. But Ken is like a midwife, I mean that's how he was with me. He sits there very close to you, slightly below you, and it's like he's delivering a baby when he's trying to get you to say things. He emotes encouragement. He emotes enthusiasm. He smiles at you. He looks very intently in your eyes. He nods. He's a very physically responsive interviewer. I remember my image [of him] in coming away from this interview was one of a midwife, and I think the reason that Ken has gotten such emotionally varied interviews from people over the years is because he is like that as an interviewer. What he is conveying to you through his style is that "I'm going to try to bring out the very best that you have to say and we're going to stay here until you say it, and we're all in your corner," and his body language and eye contact as well as his questions all covey that.[48]

The historical narrative evolved in a deliberate and discernible fashion. The first two drafts of the script and the initial rough cut contain vast amounts of background material on radio aesthetics, programming trends, and the emergence of the broadcasting industry, advertising, and networking; the popularity of major personalities and the habits of the listening public, as well as the ever-looming presence of television, which arrives suddenly and dramatically toward the end of Act 3. The fine tuning that ensued in the "Coda" of *Empire of the Air,* however, is instructive about the development of the documentary as a whole.

In Paul Barnes's preliminary rough cut, multiple commentaries by Norman Corwin, Garrison Keillor, and Robert Saudek comprised a nearly 10-minute finale stressing the legacy of Armstrong, the many stylistic charms of radio, and television's impending eclipse of the older medium.[49] Barnes's final version of the "Coda" in June 1991 is extensively honed to just one minute three seconds, composed mostly of an amusing anecdote by Corwin (whom Burns refers to as "the Greek chorus in the film"[50]): "A child was asked whether he preferred radio to television, and he said, 'Radio!' and the father said, 'Why?' and the child answered, 'Because the pictures are better.'" This sunny counterpoint coming less than seven minutes after Armstrong's tragic suicide, once again, underscores the uneasy relationship in *Empire of the Air* between the tragic saga of the radio pioneers and the much more light-hearted remembrances involving the golden age of radio and its attendant cultural impact.

Empire of the Air was originally planned as a 90-minute documentary, which is almost precisely the screen time currently devoted to the de Forest, Armstrong, and Sarnoff triple biography in the final 120-minute print. When a three-hour version of the documentary was reduced to just two hours in the early to mid summer of 1991, much of what was eliminated were the broader environmental details concerning the social construction of radio broadcasting. The scripting and editing decisions, therefore, were principally made to serve the "Radio Pioneers" storyline; and constructing a tight, lean narrative flow was the main reason why the backstage drama between de Forest, Armstrong, and Sarnoff was thrust so prominently front and center, while the contextual matters were pared down to approximately thirty minutes in the released version of *Empire of the Air*. As Ken Burns remembers, "We originally thought *Empire* would be a single 90-minute film and once considered three single-hour programs. Yet for dramatic purposes, we did not want to divide it, so we opted for a very long and difficult format [120 minutes]—one that you really have to strap on your hiking boots for, and I'm delighted that we're pushing the boundaries of attention."[51] Many of the rougher edges and evident inconsistencies in the personalities of de Forest, Armstrong, and Sarnoff were similarly reworked over the course of scripting and assembly. At the outset, Burns had recommended a wide array of contradictory descriptions from the various interviewees for Geoff Ward's consideration, including Marconi as "the greatest"; de Forest as "the Thomas Edison of the radio world," "a great inventor," "I have a lot of respect for [him]," "glamorous . . . [and] . . . good looking," "sweet, gentle . . . [and] . . . very unpretentious"; Armstrong as "a brilliant but difficult man," "stubborn . . . and inflexibl[e]," "a very lonely man," "despondent," "paranoi[d]," even motivated by money, and "I think the only one who killed Armstrong was himself"; and Sarnoff as "the real genius," "a visionary," "a dreamer," "an extraordinary person of great breadth," and "a good father."[52]

The earlier drafts by Ward were far less complimentary of Armstrong, for instance, and provided many more sympathetic insights into de Forest and Sarnoff. Burns's written responses to these complexities of character were consistently confirming and enthusiastic.[53] Advice was, moreover, solicited from Barnouw, Bilby, and Kelley, while Roloson and Douglas reviewed the initial version of the screenplay and responded to a videotape of the rough cut which was mailed to them for their professional opinions. By the fourth or fifth drafts of the script and corresponding work print edits, however, the portraits become less ambiguous as the many overlapping interpretations of the radio pioneers were either cut down (as had occurred earlier with the material on radio programming and culture) or were reconstituted into the sharply defined narrative that now decisively asserts Armstrong as the man most responsible for the invention of radio.

Burns and his colleagues, in essence, expressed their enthusiasms for the protagonist by slowly enhancing his position step-by-step during the editing process. They likewise gradually diminished the relative importance of the two antagonists in the course of their rebalancing the historical record in favor of Armstrong. The admittedly flawed de Forest, for instance, was in actuality far more competent than the amusingly inept foil presented in the documentary. In his filmed interview, Erik Barnouw sheds considerable light on this inventor's later shortcomings by speaking at length about the stark deprivation and loneliness of his childhood: "Lee de Forest is a person who haunts me."[54] Susan Douglas, too, clarifies in her commentary that "I don't mean to suggest that he [de Forest] wasn't a skilled technician, he was—he also happened to be a thief and a cad and a real stocksharper."[55] Kenneth Bilby offers a similarly fuller impression of Sarnoff as "an extraordinarily complex human being" who was, at once, "farsighted . . . very imposing . . . [and] affectionate."[56] The historical de Forest and Sarnoff were less foolish and venal respectively and far more complicated than their screen counterparts, just as Armstrong built on the discoveries and imaginings of others as does any innovator, no matter how gifted. As engineer Loren Jones points out in one of his outtakes: "There probably is no one father of radio."[57] After all, even *Empire of the Air* takes its name, ironically, from the metaphor that de Forest conjured up to describe his remarkably prescient notion of radio broadcasting, an insight that eventually required Armstrong's engineering genius to become a workable reality.[58]

These abridged portrayals, by and large, have much more to do with condensing the narrative from three hours to two to ensure *Empire of the Air*'s storytelling vitality and effectiveness, as well as to fit it more comfortably into a single evening of viewing, than with any inherent limitations in using a biographical approach to cover this particular subject. The personal ambitions, passions, jealousies, and broken friendships among de Forest, Armstrong, and Sarnoff did, in fact, profoundly affect the history and development of early radio. In addition, the intimate nature of television watching in general, taking place in the privacy of one's own home on a relatively small screen, is especially well suited to this method of historical presentation. Burns and his creative team seamlessly weave together a highly organic, interlocking plot structure which provides a legitimate, albeit abbreviated, interpretation of three of the most seminal radio pioneers who today are largely unknown to most viewers. As Erik Barnouw concludes in retrospect: "I liked it [*Empire of the Air*] very much. I saw it several times and I was very pleased with it . . . I am very enthusiastic about television as a medium for history. I think film presentations of history are very important and they make an enormous impression as an introduction. It makes [history] more vivid."[59]

Empire of the Air, overall, is a model example of how to combine the popular and scholarly traditions by concurrently translating a historical subject into different though mutually beneficial media. Most books published and released in tandem with major motion pictures or television series are commercial ventures which are typically heavy on promotional photographs and light on text and analysis. The written version of *Empire of the Air: The Men Who Made Radio*, in contrast, stands as an ideal complement and cross-reference work to its film counterpart. Many of the pertinent issues raised in the documentary, such as the progressive moves from wireless telegraphy to wireless telephony to radio broadcasting or the many interlocking court cases related to frequency modulation, are described and evaluated to a much greater degree in the printed account. Tom Lewis, too, freely developed his portraits of de Forest, Armstrong, and Sarnoff in far more depth in his published volume, *Empire of the Air*, even integrating many of the fascinating personal details that were eventually edited out of the documentary in its final stages. As Lewis explains, "This is a serious book. It's thought of as a companion book to the film when, in fact, the TV show should really be thought of as a companion to the book, but, of course, nobody sells TV shows that way."[60]

The history and development of radio, in the end, is a far more complex and sweeping enterprise than any one television documentary or book can adequately address and explain all by itself, no matter what historiographic approach is used. *Empire of the Air: The Men Who Made Radio* in its various forms makes a significant contribution to the existing literature as both popular and scholarly history, mixing well with the rapidly growing number of related books that investigate radio first and foremost from the contexts of culture, commerce, and politics.[61] As one sign of the standing of *Empire of the Air*, a media professor noted shortly after the film's release, "none of the nearly fifty published reviews [in newspapers and magazines] expressed a single reservation."[62] Morgan Wesson remembers that "the press was sharpening its knives after *The Civil War* and then he [Burns] comes back with *Empire of the Air*."[63]

Most academic critics reacted similarly, although a few scholarly reviewers took particular issue with the decision to limit the men who made radio to only de Forest, Armstrong, and Sarnoff.

One media historian who admiringly noted "the exquisite handling of the tale of these three men by Ken Burns, the video's producer, and by Lewis," for example, also pointed out "to view Marconi as somehow unconnected to the development of radio as a mass medium is a little like saying that Thomas Edison should not be discussed as the progenitor of sound recording technology because he intended to pursue it as a business dictation device."[64] Another media historian made a case for Marconi, Fessenden, and the desires of the mass audience as the three most "important component[s] in the history of radio" after speculating that "the programme [Empire of the Air] has been widely viewed in classrooms and elsewhere, and shows much promise of becoming an (if not the) accepted version of U.S. radio history."[65] Too often, made-for-TV histories are hastily misperceived as the last word on any given subject, simply because of the unprecedented power and influence of television as a medium. Rather than being definitive, the film version of Empire of the Air is best understood as a dramatic alternative to the many published histories of radio. This documentary revivifies three of the most instrumental pioneers in broadcasting's earliest years for television and classroom audiences, spurring some of these viewers to pursue this new-found historical interest beyond the screen and into other forms of popular and professional history. A cultural historian aptly concluded his review on this very point: "Paired with the work of scholars such as Daniel Czitrom or Susan Smulyan or supplemented with a thorough discussion of the issues presented and omitted from it, the film is an engaging and effective teaching tool."[66]

A HINT OF THINGS TO COME

I feel very strongly that one [adapts stories to] different media. And I was looking for one medium for the book, another medium for the film . . . [and] a radio program that was based on the book.

—Tom Lewis, 1999[67]

Empire of the Air was initially slated to be delivered to WETA-TV during the first week of January 1991 with the understanding that the premiere PBS telecast would occur later that April. The bustle and excitement created by The Civil War, however, pushed the original work schedule to the late summer and early fall of 1991, when the closing phases of assembly culminated with the final sound mixing, the title and optical work, and the fine tuning of several successive answer prints. A master copy and a number of release prints were struck in October which formally marked the end of the production process. HarperCollins also published Tom Lewis's hardcover version of *Empire of the Air: The Men Who Made Radio* in late October under the Edward Burlingame imprint, thus making it available to bookstores nationwide just before the start of the holiday season.[68] Tom Lewis was also consulting with a radio producer from WETA-FM, Mary Beth Kirchner, and one of the founding members of the Firesign Theatre, David Ossman, who together were developing a radio adaptation of the radio pioneers story.[69] Yet a third variation of *Empire of the Air* was nearly ready to debut on American Public Radio in early February 1992.

The two-hour film aired on Wednesday, January 29, 1992, garnering a 3.8 rating which was nearly twice the PBS prime-time average at the time.[70] *Empire of the Air* attracted 6.32 million viewers with that inaugural showing, further reinforcing Burns's standing and reputation as one of public television's most watched, celebrated, and accomplished producer-directors. Ken Burns, most significantly, was now assuming a far greater role as the chief executive and guiding force at American Documentaries, Incorporated to go along with his usual directorial duties. He was juggling much larger opportunities and obligations than ever before following the phenomenal success of *The Civil War*. Burns's home base in Walpole, New Hampshire, was quickly transforming from what was essentially a small family operation to what would soon become a much larger and more extensive independent production facility.

As a result, Burns increasingly relied on his talented group of core associates who basically continued as his ongoing creative team, especially Geoff Ward, Paul Barnes, and Buddy Squires. He also collaborated regularly with an assortment of coproducers on multiple projects in various stages of development. He and Tom Lewis clearly shared a common vision of how to realize *Empire of the Air* on film, a perspective which Morgan Wesson eagerly adopted and helped them implement once he joined the effort. The active production phase of *Baseball,* moreover, was already in full swing during the major league season of 1991. Burns was, thus, deeply involved in producing and directing this extended miniseries with coproducer Lynn Novick; and he was preparing *The West* with Stephen Ives who would act as his on-the-line producer-director as he assumed the role of executive producer on yet another sweeping and ambitious endeavor.

The debut telecast of *Empire of the Air* ultimately provided a coattail effect for the companion radio drama a week later. Like the made-for-television history before it, this 90-minute sound version of *Empire of the Air: The Men Who Made*

Radio was also a dramatic adaptation of Tom Lewis's now best-selling book. Mary Beth Kirchner had originally proposed doing a radio play to Lewis, who was instantly intrigued with the idea. She then enlisted David Ossman, who had just written and directed the Grammy-nominated *War of the Worlds 50th Anniversary Production* for American Public Radio.[71] He brought his vast experience and expertise as an FM radio pioneer to what became a richly textured audio rendition of the "Radio Pioneers" plot line, interspersed with numerous recreations of historic broadcasts using authentic wireless telephony and telegraphy equipment and techniques from the period. The result is a more historically detailed, word-based drama than the film, employing an array of talented performers as its own chorus of voices, including Steve Allen as the narrator, David Ogden Stiers as Armstrong, John Randolph as Lee de Forest, Harris Yulin as David Sarnoff, Ed Asner, John Astin, Bonnie Bedelia, and even reprising Norman Corwin from the television version.[72]

Along with the television special, General Motors was, in the end, the corporate underwriter for the public radio version of *Empire of the Air;* both productions were funded as part of the General Motors Mark of Excellence series. The budget for the historical film documentary had eventually risen from the initial projection of $627,334 to just over $800,000. General Motors agreed to offset the difference as well as pay for additional promotions and educational outreach materials. Tom Lewis recalls that "GM put in a fistful of money and became the corporate sponsor, but that didn't happen until after *The Civil War* went up and that, of course, was an enormous success."[73]

The Civil War also paved the way for *Empire of the Air*'s ensuing distribution agreements for classroom use and the home video market. Like its predecessor, *Empire of the Air* was licensed to educators for off-air taping during both its debut and its public television rebroadcast on March 31, 1993; video distribution rights were sold to Pacific Arts Corporation, a small independent firm which established an exclusive arrangement with PBS Video in 1990. Pacific Arts was summarily dropped by PBS in October 1993, however, for chronically owing back video royalties and unpaid advance guarantees. PBS eventually signed a distribution agreement with Turner Home Entertainment in April 1994: "A dozen titles have been selected, among them Burns' 'Empire of the Air,' a history of broadcasting aired earlier, and 'The Civil War,' making a return engagement on cassette in June [1994]."[74] Lawyers for Burns's companies, American Documentaries, Inc., and Radio Pioneers Film Project, Inc., moreover, filed suit in U.S. District Court in Los Angeles on October 7, 1994. "The lawsuit claim[ed] that Pacific Arts ha[d] failed to make any royalty payments on 'Empire of the Air' since April 1993." It also cited a "$1 million unpaid balance for 'The Civil War.'"[75]

Clearly the market had fundamentally changed for made-for-television histories. Turner Home Entertainment, for its part, began packaging *Empire of the Air* after 1996 in a collection with six one- to two-hour documentaries by Burns from the 1980s covering an assortment of subjects from American history.[76] "Ken Burns has become a brand name in himself," announced Vito Mandate, director of strategic marketing at Turner, "consumers have come to think of him in a unique way."[77] Nonfiction histories were, additionally, occupying prime-time slots all over the TV dial, on commercial and public networks, from corporate and independent producers. Burns had now emerged as the signature figure for this much larger programming trend. His two-decade long, single-minded focus on producing historical documentaries for television, a minor programming niche at best throughout the 1980s, had finally and improbably gone mainstream by the mid-1990s. And Ken Burns was already looking ahead toward his next big project. The final image of *Empire of the Air* tellingly features a lightning streak crackling across the evening sky, accompanied by the sound of a telegraph typing out in Morse code: "B-a-s-e-b-a-l-l-i-s-n-e-x-t."

CHAPTER FIVE

A Whole New Ball Game: *Baseball* (1994) and *The West* (1996) as Event TV

A LEAGUE OF HIS OWN

I took a vow of poverty and anonymity when I became a documentary filmmaker, and neither of them came true.

—Ken Burns, 1999[1]

The initial idea for *Baseball* was spawned in a conversation over drinks during the early stages of the Civil War Film Project. "I was in a bar in Georgetown in 1985 with a friend and producer, Mike Hill," remembers Ken Burns. "He mentioned baseball and all the bells went off inside. The whole time I was working on *The Civil War,* I knew I'd do *Baseball.*"[2] Hill's suggestion was that the national pastime would serve as the ideal vehicle for understanding and representing on film the kind of country that the United States had become in the century following its seminal conflict. Burns next shared his thoughts about *Baseball* with the well-known historian William Leuchtenberg at the Organization of American Historians (OAH) annual conference the succeeding year, when the producer-director was there to accept his second Erik Barnouw Prize for the Outstanding Documentary Film in American history for *Huey Long.* Burns was "concerned that baseball might seem too frivolous a subject."[3] Leuchtenburg, who had already advised the filmmaker on *Huey Long* and was acting in a similar capacity on *The Civil War,* assured him that "baseball [was indeed] a challenging subject [and that the game was] just now enjoying the kind of scholarly attention that would benefit [his] work."[4] The University of North Carolina professor "was also an avid fan and student of the game." Thus, he agreed to write a concise outline on the history and significance of baseball as a starting point for "discussions about the architecture and thematic focus of the series" that ensued between Burns and Geoff Ward, who also signed on as head writer for the series in 1987.[5]

Preoccupied with finishing *The Civil War* as well as completing *Thomas Hart Benton* and *The Congress,* Burns did not make another major decision involving *Baseball* until the spring of 1990, when he invited Lynn Novick to share primary responsibility with him for producing what, at the time, was envisioned as a five-part, ten-hour series. Novick had been working in film and television production since 1986, mostly at Thirteen/WNET in New York, following her graduation from Yale University in American Studies. Before joining Florentine Films/American Documentaries, she worked for two years with Bill Moyers as an associate producer and researcher on several episodes of his *World of Ideas* (1988) series and on his widely acclaimed and popular *The Power of Myth* (1988) featuring Joseph Campbell. After these experiences, she joined Burns as his associate producer for postproduction on *The Civil War* and immediately impressed him with her drive, dependability, and ability to work with a wide variety of people. She at once began the myriad tasks necessary for researching and preparing *Baseball,* continuing throughout the remainder of 1990 as Burns devoted the vast majority of his time to promoting *The Civil War* and shooting and editing *Empire of the Air: The Men Who Made Radio.*

When the "Radio Pioneers" film project crew traveled to Tallahassee, Florida, in the fall of 1990 to interview Red Barber, baseball's veteran radio and television announcer, Burns framed his questions in such a way as to also conduct his first official on-camera exchange for *Baseball.* The staff at Florentine Films had already received the good news in July that the Corporation for Public Broadcasting had awarded them $1 million from their challenge fund, while frequent collaborator Stephen Ives had been placed in charge of assembling a much larger proposal to the National Endowment for the Humanities for $1.5 million, which was submitted in March 1991. General Motors agreed to commit $1.5 million toward the production of *Baseball* on the heels of *The Civil War*'s success in September 1990; the automaker likewise committed a similar amount to promote and market the series and to provide an extensive educational outreach component again. NEH notified the filmmakers in August that their $1.5 million request would be awarded in full as two private foundations, the Pew Charitable Trusts and the Arthur Vining Davis Foundation, also expressed keen interest in underwriting portions of the project. By the time formal editing began in November 1991, Burns and company had met and slightly exceeded their original budget projection of $4,333,318.[6] Although a good deal of time and effort had been invested in soliciting money for *Baseball*, the extraordinarily positive reaction to *The Civil War* and Ken Burns's heightened profile clearly accelerated the process and made it much less arduous than ever before. The fundraising struggles for *Brooklyn Bridge*, or even *Empire of the Air*, were now behind the producer-director, at least for the foreseeable future.

The relative ease of financing *Baseball* encouraged Burns and his creative team to expand their conceptual ambitions for the series. "After the first six months of editing, which were spent preparing the rough assembly of the film," recounts coproducer Lynn Novick, "we determined the story we wanted to tell could not be told in ten hours."[7] As a consequence, the filmmakers resolved to extend *Baseball* to nine episodes with each installment lasting approximately two hours. They revised their budget accordingly and in January of 1993 received additional grants totaling $1 million from General Motors, the Corporation for Public Broadcasting, and the Public Broadcasting Service.[8] *Baseball*'s final budget reached $7,637,000 with the addition of more NEH and private foundation support, making it public television's most expensive and "longest single subject program ever."[9]

Producer-director Ken Burns was, in effect, redefining the possibilities and expectations for PBS. With the unprecedented popular response accorded to *The Civil War*, public television now began to cultivate certain special blockbuster-size events as never before, appreciably raising the bar with *Baseball*, to parallel their traditional mission of providing a viable noncommercial alternative to the major private sector networks. At the same time, newly created cable services, such as Arts & Entertainment (A&E), Discovery, and the Learning Channel, were clearly making inroads into PBS's programming distinctiveness. In an era of waning governmental support, public broadcasting executives were increasingly under pressure from Congress to justify the system's very existence. *The Civil War* offered an answer for many. Ward Chamberlin, president of WETA in Washington during the production and debut of *The Civil War* and a former vice president at PBS remembers, "[t]his is what we've all hoped public television could be. For me, after 20 years in the business, it's the most crowning achievement one could have . . . Congress will continue to support us to the extent they have. One senator told me, 'Now we have something we can point to instead of just *Sesame Street* and the *MacNeil/Lehrer Newshour*.'"[10]

Ken Burns was now playing in a league of his own as a documentarian. The irony of Burns's triumph at PBS, according to James Day, "was the fact that public television's greatest programming success was not, as one might expect, produced by the system's complex bureaucracy and its billion dollar kitty. *The Civil War* was a product of the community of independent producers working outside the system, making the kinds of shows they most want to make, struggling to find the dollars to fund them, and hoping that the finished product will find its audience through acceptance on the public-television system."[11] Burns, in this regard, is merely one of literally hundreds of self-supporting filmmakers working on the periphery of the Public Broadcasting Service. "Very few, however, have had his success as an independent entrepreneur . . ." Day continues. "He has long since learned what every independent ultimately comes to know: that creativity alone is meaningless without the patience and skills to find the dollars to make it happen."[12]

Burns, furthermore, was feeling his way in terms of how entrepreneurial he needed to be after the release of *The Civil War*. As the chief executive at Florentine Films/American Documentaries, he ventured into merchandising a wide assortment of ancillary products for the first time in his career with *Baseball*. Burns had learned a hard lesson with *The Civil War*. When that miniseries "became an unexpected sensation in 1990 . . . [p]eople wanted to do plates and mugs," explains the filmmaker, "I thought it was crass." He watched as unauthorized "horrible knockoffs flooded the market."[13] Plans to circumvent such commercial pirating with *Baseball* began with preliminary discussions as early as 1992 to incorporate a new subsidiary attached to his company, specifically for the purpose of licensing and selling product tie-ins.

Burns entered into a formal agreement in the fall of 1993 as the majority shareholder of an affiliated corporation called Baseball Licensing International with Carbone, Smolan, and Shade Associates, a New York-based firm designated to prospect and field incoming merchandising proposals of all kinds, with John Thorn, a widely published expert on the sport and founding editor of *Total Baseball* (the most thorough statistical record of the game ever published) who was now extending his miniseries consulting duties to the choice of commodity tie-ins, and with the attending lawyers in New York and New Hampshire.[14] According to the contract, "Ken Burns will have final creative decision-making authority on all matters that affect the company."[15] The independent filmmaker was now exercising his judgment in a wholly new area by making literally dozens of sales determinations with his business associates.

All told, Burns, Thorn, and Novick reviewed more than 40 product proposals which were arranged across four broad categories, including five different sets of trading cards; wall, desk, and pocket calendars; an array of publications directed at adults and children; as well as 20 consumer items ranging from bats and balls to lunch boxes and shower curtains.[16] Burns and his colleagues did, in fact, veto more than half of these suggested tie-ins; still, no documentary film in history has ever been released to the public with quite the level of commercial fanfare as *Baseball.* Florentine Films/American Documentaries/Baseball Licensing International, the Public Broadcasting Service, and General Motors joined forces with Turner Home Entertainment, Alfred A. Knopf/Random House, and *Sporting News* to enact an advertising, marketing, and retail campaign comparable in size and sophistication to those regularly mobilized by the major Hollywood studios and commercial television networks for their motion picture and series premieres. The fulcrum of all the synergistic merchandising strategies, moreover, was the talented producer-director turned able celebrity spokesperson—Ken Burns.

Many of the GM-related executives most responsible for publicizing *The Civil War* recognized straightaway in 1990 that Burns was playing a pivotal role in maximizing the success of his documentary film. The filmmaker's boyish good looks coupled with his outspoken passion for history and his heartfelt belief in his own work made him an appealing emissary for the series in his many encounters with the press and potential viewers nationwide. "Ken did a lot of interviews and was great copy," reported Richard Prince of N. W. Ayer who handled the GM account for the advertising agency, "he also mentioned the fact that GM helped to fund the project whenever he could."[17] Owen Comora, head of Owen Comora Associates, the public relations firm hired by GM to promote *The Civil War* and other Burns television specials up through *Frank Lloyd Wright,* formed a similar impression: "Even though [Ken] was on the road, he would call in every day for his interview schedule. He said he would do whatever needed to be done. Plus, I've never met anyone as quotable."[18] Burns willingly accepted most every assignment he was offered to publicize *The Civil War;* he likewise committed himself to an even more extensive promotional schedule for *Baseball.*

Ken Burns's personal appearances for his newest extended miniseries began on Thursday, May 26, 1994, with a Baseball Hall of Fame dinner in New York City and continued nearly nonstop through Wednesday, September 21, with a private screening and celebrity fundraising party in Los Angeles to benefit the Negro League Museum and the Jackie Robinson Foundation, held in conjunction with the *Sporting News* and the *Los Angeles Times.* For the better part of four months, Burns crisscrossed the country a dozen times, attending a dizzying array of events created and sponsored by PBS, General Motors, *Sporting News,* major league baseball, Turner Home Entertainment (who distributed the accompanying videotape), and Alfred A. Knopf (who published the companion book), among other associated companies, subsidiaries, and institutions.

During the long weekend of June 2 through 5, for example, the producer-director was a featured speaker at the PBS Annual Convention in Orlando, Florida, for the more than 1,100 public television representatives attending from around the country. While introducing a brief videotaped preview of *Baseball*, Burns also listed a broad selection of pledge items related to the series which could be utilized in future on-air drives, such as "CDs, leather jackets, mugs, books, T-shirts, and even polo shirts with a little GM logo printed discreetly beneath the series' title."[19] The filmmaker, sporting a recently grown beard and a navy blue tie with a pattern of white baseballs with red seams, also signed autographs and posed for photographs with a long line of public broadcasters. He was enthusiastically embraced by the PBS faithful as one of the network's biggest stars. His increasing visibility with the general public and congressional lawmakers, moreover, induced one station executive to remark that "Ken Burns is [now] the poster boy of PBS."[20]

Other highlights on Burns's nationwide publicity tour for *Baseball* included the *Sporting News* Roundtable at the July 12 All-Star Game in Pittsburgh with Hank Aaron, Buck O'Neil, the 82-year-old former Negro League ballplayer and manager who is a featured commentator in the series, and with several sportswriters; a 15-market PBS swing stopping at various major league ballparks around the country where Burns met with a succession of local dignitaries and journalists, held dozens of press conferences, and, more often than not, threw out the first pitch in opening festivities preceding each game; and a special Wednesday, July 27 screening and reception at the Television Academy of Arts and Sciences in Los Angeles (*Baseball* eventually won a 1995 Emmy for Best Informational Series). There was also a Saturday, September 10 screening and reception at the National Gallery in Washington followed by a picnic for several hundred on the White House lawn hosted by President Clinton; a Monday, September 12 testimony before the House Subcommittee on Telecommunications and Finance concerning governmental appropriations for the Corporation for Public Broadcasting; a Tuesday, September 13 *Baseball* reception for Congress sponsored by the National Endowment for the Humanities; and spot presentations and interviews on ABC's *Nightline,* the *CNN Morning News,* public TV's *Charlie Rose Show,* ABC's *Good Morning America,* NPR's *Morning Edition,* among many other national and local television and radio programs.[21]

Baseball's corporate sponsor, General Motors, and Burns's other major commercial partners—Turner Home Entertainment, Alfred A. Knopf/Random

House, and Elektra Nonesuch—also orchestrated elaborate marketing campaigns of their own to coincide with the Sunday, September 18 public television premiere of the miniseries. General Motors, for instance, underwrote a 30-minute promotional documentary entitled *The Making of Baseball* (1994), which was produced by New Hampshire Public Television and then made available to PBS affiliates around the country for telecast beginning on July 4 and continuing throughout the rest of the summer leading up to the program's mid-September debut. GM also hosted a nationwide student essay contest on "Baseball, Me, My Family & Friends," in which the grand prize winners—one boy and one girl—were each awarded $10,000 in college scholarship money and all the regional finalists received $1,000 Series EE U.S. Savings Bonds.[22] In late August, GM mailed out 40,000 teacher guides called "*Baseball*—Lessons of the Game," continuing their comprehensive educational outreach program as a major component of support for Burns's work. Teacher response was so "enthusiastic" to this instructional kit of lesson plans and classroom activities that "100,000 additional copies of the guide [were soon] requested" after the initial mailing was announced.[23]

Turner Home Entertainment, for its part, marshaled a $25 million advertising and marketing campaign for the release of the videotape version of *Baseball* to coincide with the series' televised premiere.[24] The home video producer-distributor placed print ads in many mass-circulation magazines, including *Newsweek, People,* and *Sports Illustrated;* it purchased 15- and 30-second spots on the A&E, Discovery, ESPN, and USA cable networks; and it ran additional commercials for the complete set (retailing at $179.98) and individual episodes (at $24.98 each) on Turner's own CNN, CNN Headline News, WTBS, and TNT. "What's unusual from our standpoint," explained Russell Kelban, Turner Home Entertainment's vice president for marketing, "is to go day-and-date with PBS. 'The Civil War' was basically released [on videotape] a year after [the debut], and there was incredible success with that product. So we can only imagine how well we're going to do with a product that is much more mass-oriented, and releasing it day-and-date."[25]

On September 7, Alfred A. Knopf delivered 500,000 first edition hardcover copies of its $60 companion volume, *Baseball: An Illustrated History,* while its subsidiary, Random House, released its 220-minute audio-cassette and CD versions of the book (read by Ken Burns himself) and three related Random House titles—*25 Great Moments, Shadow Ball: The History of the Negro Leagues,* and *Who Invented the Game*—targeted more for the young adult and juvenile reader.[26] Elektra Nonesuch completed the complement of entertainment companies saturating the market with mediated tie-ins related to *Baseball* by releasing its soundtrack on September 6, having sold an estimated half-million copies of the companion audiotape and CD for *The Civil War*. "We're doing extensive merchandising on this, including a special retail counter piece," described Peter Clancy, Elektra Nonesuch vice president for marketing and creative services. "We have ambitious co-op plans for pricing and positioning, starting at the end of September and taking us through Christmas."[27] On Sunday, December 4, Burns even made a follow-up two-hour appearance on the Home Shopping Network to help promote an assortment of related tie-ins at the outset of the holiday season.

The full extent and aggressiveness of the advertising and marketing campaign surrounding *Baseball* understandably resulted in a modest backlash against the upcoming program as well as against Burns for the first time in his career, although there also was an evident and expected rise in positive press coverage and increased public awareness of the miniseries. *The Making of Baseball*, for instance, was panned by *Washington Post* columnist, Jonathan Yardley, who wrote, "From where I sat, [*Baseball* looks] about as enticing as a striptease by the circus fat lady."[28] Burns responded defensively, pointing out that the critic "didn't see one second of my film," while also conceding that the teaser was "a very poor half-hour film," giving "the impression I didn't follow my consultants."[29] A personality profile in the September 1994 issue of *Life* was another case in point. Midway through the piece, author Robert Sullivan inserted an aside which transparently disclosed his impressions of the filmmaker trying to promote himself and his series:

> You get used to the dramatics when you talk with Burns . . . Consider: When asked if there's a thread running through his work, Burns rephrases the question and continues thusly: "What is it I'm attracted to? That's another way of asking, Who am I? That's a really big question. I'm fascinated by the history of this country and what makes us us. I'm working in a new Homeric form in which you can sing the epic verses of our people." As you listen, you realize he isn't kidding. He's a believer—in good and evil, right and wrong, in himself and in his talents. When he likens himself to Homer, there is that moment when you flinch. Then you realize: He believes it! Suddenly he smiles, as he often does when the going gets heavy. It's a charming, disarming smile. Don't think me too ponderous or pompous, Burns seems to say. I know what that sounded like. But think it over. It's true.[30]

The press's overall reaction to Burns was growing far more complicated than it generally had been during the release of *The Civil War* nearly four years earlier.

He no longer appeared to be just this refreshingly young, sincere, and idealistic independent producer-director who somehow materialized unexpectedly from small-town America to match, if not beat, the rich and powerful networks at their own game. Part of Burns's original appeal in his rise to fame (and fortune) was that he and *The Civil War* seemed to be something set apart from the overhyped, overcommercialized environment which typifies most of television. Some of his difference was now fading away amid all the hoopla and related merchandising associated with *Baseball*. This change was brought home to him firsthand when an assistant offered a dozen baseball greats (or their heirs) between $250 and $500 to license the use of their photographs in wall and desk calendars from Random House.[31] Counteroffers from these former players or their agents soon shot back as high as $10,000.[32] "You lose the anonymity that was your bargaining power," lamented Burns at the time, "It's been a real eye-opener."[33]

In the three years in which *Baseball* became Ken Burns's main professional focus—roughly from the summer of 1991 when *Empire of the Air* was winding down and principal shooting began on the 18-hour 30-minute miniseries until its nine-night debut (September 18-22 and 25-28, 1994) on public television— virtually "everything about [his] life changed except the kind of father that [he is] to [his] daughters and the attitude and approach [he] bring[s] to [his] work."[34] During this period, Burns grew increasingly more prosperous ("a rare bird within the documentary breed" according to *Life*), mostly as a result of the residuals from *The Civil War*'s companion book and videotape set, along with the many paid public speaking engagements he had accepted.[35] *Baseball* too became a popular success, as 28 million viewers tuned into its premiere, while the accompanying videotape and print versions were still selling briskly more than a year after the initial telecast.[36] Burns felt progressively more unsettled about all the commercial activity, however, and elected in the end to redirect a large portion of the income to a handful of worthy causes. "I started with plans for a full-scale, profit-making enterprise," he revealed, "but that made me uneasy. I decided to take my share and pour it into non-profit organizations."[37] He, in this way, donated all his proceeds, "except from videotapes and a book," to the Negro League Museum in Kansas City; Harlem RBI, an organization "which gives inner-city youth the opportunity to play baseball"; the Baseball Hall of Fame in Cooperstown, New York; "a fund for impoverished former baseball players"; the National Council of History; "and a fund to support deserving young filmmakers."[38] Burns, most remarkably, fully reimbursed the National Endowment for the Humanities (as he had done previously for its financial backing of *The Civil War*) for what ultimately turned out to be $2 million in funding support for *Baseball*.[39]

Ken Burns's multiyear commitment to *Baseball* was, ultimately, a time of unexpected and tumultuous highs and lows for him both professionally and personally. He had often contended during the exhilarating aftermath of *The Civil War* that "our family life is held together by Amy with an understanding I don't deserve."[40] By 1993, though, his 11-year marriage "to his college sweetheart unraveled," as he and his soon to be ex-spouse agreed to separate, although "both continue [even today] to live in Walpole and share custody of their children."[41] Burns characteristically "threw himself ever deeper into *Baseball*" in reaction to the breakup.[42] Just like the ever-expanding ambition and length of the project that now preoccupied most of his time and attention, Burns's life and career as a documentary filmmaker had developed commercially and creatively into a whole new ball game.

MORE THAN A GAME

I think for those of us that love something that's slower-paced, that has a history to it, that has memories attached to it, baseball's perfect.
—*Doris Kearns Goodwin, 1997*[43]

By mid-September 1994, expectations were running high for *Baseball*. Much of this anticipation had to do with the record-setting success of *The Civil War*, along with the combined effects of *Baseball*'s advertising and marketing campaign and the apparent aspirations of the series' producer-director, Ken Burns. His argument, articulated in hundreds of interviews across the country,

> was that if you wished to know the country that the Civil War made us, you could find no better vehicle to find that out than studying our national pastime. This is the story of race, the age-old tensions between labor and management, the history of immigration and assimilation, the exclusion of women, the rise of popular media, advertising, and popular culture, the growth and decay of cities; all of these broad themes were on display in our national pastime and in near perfect collision throughout the history of the game.[44]

Almost on cue to prove his point, major league baseball inexplicably self-destructed before the nation's eyes on September 12 as the Major League Players Association went out on a season-ending strike over a salary cap and team revenue-sharing proposal demanded by the owners. The bitter spectacle of a protracted labor-management dispute, seemingly impervious to any kind of federal mediation, replaced the World Series for the first time since 1904. *Baseball*'s September 18 premiere was scheduled by PBS programmers to coincide perfectly with the usual peak period of the American and National Leagues' pennant races. Burns's miniseries, instead, was the only game available to anyone. "On one hand, people say, 'Well, it'll be great. You'll have great ratings, because no one will be able to watch baseball on TV if the strike goes on three more weeks," explained coproducer Lynn Novick, "But we're disappointed, because people will be feeling negative about baseball."[45] The epic sweep of *Baseball* with its bittersweet culmination in the last episode entitled "Ninth Inning: Home (1970-Present)" in many ways took on the unintended feel of an elegy for the national pastime.

Baseball covers more than 150 years of America's cultural past, from the make-believe legend surrounding Abner Doubleday's supposed founding of the sport in Cooperstown, New York, in 1839 to today's era of big money, fading traditions, and the game's uncertain future fueled by internal dissension and greed as well as the ever-growing competition from football and basketball. Burns and his creative team structure their 18-hour 30-minute storyline chronologically, beginning each of the nine episodes, or "innings," with a historical review of the time period using a technique they call framing. In the "Second Inning: Something Like a War (1900-1910)," for example, John Chancellor, the narrator, provides the appropriate background information, ending with an allusion to a Yankee hall-of-famer as a way to shift the narrative focus back to baseball.

> Between 1900 and 1910, there were revolutions in China and Central America, and war between Russia and Japan. Mohandas Gandhi led a nonviolent campaign for civil rights in South Africa. Paul Cézanne and Henri de Toulouse-Lautrec died, and Georges Braque and Pablo Picasso shattered the art world. In America, Orville and Wilbur Wright flew at Kitty Hawk, Henry Ford began turning out Model Ts, and Lee de Forest patented the vacuum tube that made broadcasting possible. Geronimo and Calamity Jane and Susan B. Anthony died; Charles Lindbergh and Louis Armstrong and Lou Gehrig were born.

The filmmakers, furthermore, anchor their plot line around two of the most storied franchises of all time, the Boston Red Sox and the Brooklyn (later, Los Angeles) Dodgers, whose histories symbolize the rarity of success in the sport set against the more common scenario of daily endurance and disappointment which are emblematic of baseball as a whole. As John Chancellor summarizes in the prologue of the "First Inning: Our Game (The 1840s-1900)": "And yet the men who fail seven times out of ten are considered the game's greatest heroes." According to head writer Geoff Ward, in his preliminary 1990 outline for the series, the Boston Red Sox embody "the Eternal Puritan Quest. It is the Red Sox who struggle, year in, year out, echoing life's difficulties."[46] The scriptwriter similarly references the classic observation by long-time Boston fan Bart Giamatti, who became the major league commissioner and was forced to banish Pete Rose from the sport for gambling in 1989 and then tragically died of a heart attack a little more than a week afterward: "[Baseball] is designed to break your heart. The game begins in the spring, when everything else begins again, and it blossoms in the summer, filling the afternoons and evenings, and then as soon as the chill rains come, it stops and leaves you to face the fall alone."[47] This passage lyrically captures the long-suffering experience of being a Red Sox fan.

Boston is shown in the "Second Inning" winning the first World Series in 1903. They earn four more titles in the "Third Inning: The Faith of Fifty Million People (1910-1920)," but then a new owner trades pitching sensation and hitting phenomenon Babe Ruth to the New York Yankees in 1920 for $125,000. As John Chancellor notes in the "Fourth Inning: A National Heirloom (1920-1930)": "The Red Sox never recovered. They had won five of the first fifteen World Series; they would not even play in another series for more than a quarter century." "The Curse of the Bambino, as Red Sox fans have come to call it," wrote Geoff Ward in his early notes on the series, "has struck again and again, always preventing the beloved Red Sox from winning it all, in a litany of losses that Red Sox fans can recite with both ease and pain—from 1946 . . . 1967 . . . 1975 . . . 1978 . . . and, finally, 1986."[48] The Red Sox are presented in *Baseball* as the most tradition-bound club, embodying their New England roots. The team still plays in beautifully antiquated Fenway Park, built in 1912, typically selecting their players for power rather than speed, "watching, in some senses, the game pass them by."[49]

As the series develops, more importantly, the Red Sox are just as representative for their actions off the field as on. They are identified as the last major league franchise to integrate blacks, for instance, more than a dozen years after Jackie Robinson joined the Dodgers. Boston is likewise featured in the "Ninth Inning" as one of the two teams most responsible for bringing millions of Americans back to the game after the social turmoil of the 1960s rendered the sport increasingly irrelevant to a younger generation. The Red Sox, in this way, emerge as a worthy if overmatched opponent to Cincinnati's Big Red Machine in the 1975 World Series, the most fabled in all of *Baseball*. The stage is set for Carlton Fisk's dramatic extra-innings home run in game six, electrifying a nation and causing 75 million viewers to tune in the next night to watch Boston once again lose a heartbreaking final contest. This event holds special significance for Ken Burns, the producer-director:

> [P]art of '75 was for me realizing I didn't have to—for whatever political reasons of the war in Vietnam and counterculture—I didn't need to get rid of this game that had given me so much pleasure as a boy . . . My mother had cancer, and I was aware of it from, you know, three or four years old. I was just a little boy unable to have a childhood. I was suddenly a young man, an adult, and I had to be strong, and there were very few places where you could still be a kid. She died when I was 11, and even then that didn't stop because you had to continue being strong. And baseball was one of the few places where you could just be a kid . . . the '75 series brought me back into the game . . . I realized that I could find a certain pleasure in the game of baseball that is indescribable, and that I could take it back in and reinvest it.[50]

Ken Burns, moreover, is a Red Sox fan, obviously influencing his choice of Boston as one of the mainstays of the series. His selection of the Dodgers too, those loveable "bums" from Brooklyn, not only evokes his birthplace but provides Burns with the most appropriate site in baseball history to explore the recurring theme that animates all his work—race in America. In the "Sixth Inning: The National Pastime (1940-1950)," tennis great Arthur Ashe asserts unequivocally that "the Brooklyn Dodgers was black America's team— period." The reason for this staunch devotion by African Americans is foreshadowed during the first four episodes in several clear references to racial discrimination; still, the matter of segregation in baseball (and the United States) is brought front and center in the "Fifth Inning: Shadow Ball (1930-1940)." The subtitle for this section, shadow ball, refers to a practice originated by the barnstorming Indianapolis Clowns, an independent black ball club from that era, who "liked to warm up in pantomime, hurling an invisible ball around the infield so fast, hitting and fielding imaginary fly balls so convincingly, making close plays at first and diving catches in the outfield so dramatically that fans could not believe it was not real."[51] Shadow ball also serves as an apt metaphor in inning five for the legendary exploits of talented Negro League ballplayers such as Satchel Paige, Josh Gibson, and Cool Papa Bell, among many others who all played for decades outside the broader national awareness and limelight of the all-white major leagues.

Burns's consistent foregrounding of race in *Baseball* was a subject of some debate among his consultants. The outlook of Deborah Shattuck, a history professor at the Air Force Academy, was representative of a small minority of experts when she wrote that she could not "say enough good things about Ward's script, but I must also say I was rather disappointed in its scope. We need to broaden our vision. This script is almost 100 percent about white, professional baseball, with a sidelight about how blacks eventually made it into white, professional baseball."[52] The vast majority of advisors, however, held views similar to those of William Leuchtenburg, who responded: "I'm sure I'm not telling you anything you don't already know, but there's much too much on Negro baseball. I also suggest that we take a greater critical distance toward Negro baseball. I have no doubt that there were a lot of superb players. But we have to be wary of recollections of players from the time who say that the pitchers were so great that no one could hit them and that the hitters were so great that no pitcher could get them out."[53] The *New Yorker*'s Roger Angell similarly wrote that his "central complaint is the script's extraordinary emphasis on baseball's long-term segregation of black players: its deep racism, its cruelty to great numbers of wonderfully talented blacks, and the clear evidence that racist policies and attitudes are still easy to find in baseball. I disagree with none of this, of course, but the script goes on so often and at such length on this subject that it weakens the point and begins to seem obsessive."[54]

Even John Holway, a leading specialist on Negro League baseball, felt that there was "far too much emphasis on black/white relations. It is ironic that I should say this, since it is my own area of expertise, and ten years ago I probably would have been arguing vociferously that you ignored the black half of baseball history entirely, because that is what all histories until then did. But the pendulum has gone too far the other way, and in giving minute details of integration, you steal valuable air time from what else was going on in baseball, and there was a lot else."[55]

Probably the most controversial scene in the entire miniseries occurs midway through episode five when John Chancellor narrates over a Kansas City Monarchs team picture that "[b]y 1934, the world economy was in ruins and fascism was on the rise. In Germany, the National Socialists had come to power and begun to institute exclusionary laws against Jews based in part on Jim Crow statutes in the United States." The linkage of Negro League baseball, racial segregation in the United States, and the legislative actions of the Nazi party, though literally true, gave pause to many of Burns's advisors, including Gerald Early, the director of the African and Afro-American Studies program at Washington University, who commented that "even during the worst lynchings . . . it could never be said that the United States was moving to some final solution toward black people."[56] A discussion ensued at one of the consultants' meetings in Walpole in which several advisors suggested that Chancellor's remark might inadvertently imply a connection between Jim Crow and German eugenic theory, although no clear consensus was reached as to what effect the statement might ultimately have. Burns, in the end, decided to keep the comment in, underscoring his commitment that *Baseball* be a vehicle for examining the country's complicated and yet unresolved legacy of racial discrimination, among other important sociocultural issues.

The next episode, inning six, also spotlights race and revolves around *Baseball*'s foremost hero—Jackie Robinson. In a 5-minute 10-second scene entitled "Big League Material," for example, former Negro-Leaguer and story-teller extraordinaire Buck O'Neil recalls, "I saw baseball after the Black Sox scandal and everybody said—well baseball—they kinda got off baseball and here comes Babe Ruth hitting home runs—that brought it back . . . we went into another little recession in baseball, here comes the lights, that brought it back . . . we go into the war and all the good ballplayers are gone so that kinda brought it down a little—and here comes Jackie Robinson!"

Robinson, an All-American athlete at UCLA and a Negro League standout, formed a partnership with Brooklyn Dodger president and general manager Branch Rickey in October 1945, promising "not to fight back" for three years: "I'm ready to take the chance," Robinson is quoted by John Chancellor as saying, "Maybe I'm doing something for my race." Chancellor then goes on to describe how "[t]he pressure and the abuse were relentless," when Robinson joined the Dodgers's AAA farm club, the Montreal Royals, in 1946. His wife, Rachel, remembers onscreen: "I was sitting in a section where there were some rabid anti-Robinson people who were yelling at him, calling him names. I kept hoping that my body was blocking some of the sound 'cause I could see what their intent was." Nevertheless, "as the season progressed," relates Chancellor, "he was wracked with stomach pain on the brink of a nervous breakdown, but in the face of all the abuse, Robinson only played better." Montreal ended the season winning the AAA World Series as their star, Jackie Robinson, was called up by Brooklyn the next season, breaking the major league color line in 1947.

Robinson went on to a Hall of Fame career during a decade with the Dodgers when they captured six National League titles and finally, in 1955, the ever-elusive world championship against hated crosstown rivals, the New York Yankees. Broadcaster Red Barber claimed that he emerged as "the greatest gate attraction since Babe Ruth." Robinson was the National League's Rookie of the Year in 1947 and Most Valuable Player in 1949, but, even more significantly, his "grace under pressure" in "an adversarial, savage" environment, according to Gerald Early, "arguably . . . launched the American era of racial integration after World War II."[57] Then, in 1957, the unthinkable occurred: The Dodgers left

Brooklyn for Los Angeles, auguring a new era of coast-to-coast travel, television profits, and, finally, free agency for players. John Chancellor recounts offscreen how "Jackie Robinson did not go west with the Dodgers . . . weary and suffering from ten years of injuries and abuse," he decided to retire.

The elegiac arc of the over 18-hour *Baseball* narrative climaxes 25 minutes into inning nine with a 7-minute 50-second scene, "Safe at Home," which covers Robinson's early death at age 52 and his New York funeral attended by many luminaries, along with the remembrances of those who knew him as a player and a man. Red Barber maintains that "they said he died from diabetes and other things. I think he died from the load he carried." Most poignantly though, Robinson's widow, Rachel, recalls in close-up that

> at the funeral Jesse Jackson did the eulogy, and he said, "Jackie Robinson stole home and he's safe." And that, even now, is very important to me. Roger Kahn and his family came to visit me a week after Jack died and they had a blowup of Jack sliding into home base. And when you're looking for simple ways to deal with the grief, the deep, deep grief and mourning that you are feeling, you can catch onto a thing like that . . . so I carried that blowup from room to room for weeks just because . . . looking at it I knew he was safe. Nobody could hurt him again. He wouldn't hear the name calling. He would only hear the cheers and somehow I could fantasize my own little story about where he was and how he was doing and let him rest in peace.

Of the dozens of twists and turns in *Baseball,* the most enduring and best realized subplot is the one dealing with the racial aspects of the game in general and the exploits of Jackie Robinson in particular. As Burns recognized during a 1993 interview while he was immersed full-time in editing the miniseries, "Wait until you see *Baseball.* It's the climax—the Battle of Gettysburg and the Emancipation Proclamation rolled into one with Jackie Robinson."[58] The producer-director never overstates the obvious social significance of this racial drama which begins in inning one with the first formal exclusion of blacks from the sport in the 1880s and continues unabated through inning nine with the hate mail and actual death threats heaped on Henry Aaron as he pursued Babe Ruth's career record of 714 home runs in 1973-1974. The last episode also foretells the gradual decline of black attendance since the 1960s "as ballparks have shifted to the outskirts of the cities, and so has the percentage of black players as black athletes have found their way into other big-time sports."[59]

Despite the fading fortunes of the sport in inning nine, Ken Burns and his colleagues make a compelling case that *Baseball* was the one and only national pastime during the American century. They establish in the first episode that the popularity of the game has always ebbed and flowed, but never was the sport more in tune with the cultural currents of the country than during the "Seventh Inning: The Capital of Baseball (1950-1960)." With the Yankees, Giants, and Dodgers winning pennants, never before or since has one city dominated *Baseball* like New York did in the 1950s. The quality of the game took a giant leap forward as a floodgate of black talent now joined the usual white stars on diamonds across the United States. The old Negro League quickly faded away as African American players such as Larry Doby, Willie Mays, Roy Campanella, Ernie Banks, Henry Aaron, and dozens more joined Jackie Robinson in the majors over the course of the decade. These athletes reintroduced a faster, more scrambling style of play "called 'tricky baseball' [which] had largely been absent from the big leagues since Ty Cobb's day."[60]

Many reviewers of the series noted that "the finest work in *Baseball*'s rendering of the Negro League's experience is done by John J. ('Buck') O'Neil, a former Monarchs first baseman and manager. O'Neil holds a position in *Baseball* similar to Shelby Foote's in Burns's earlier film on *The Civil War:* a man whose stories and comments you always want more of."[61] Lynn Novick originally tracked down O'Neil after learning at the National Baseball Hall of Fame in Cooperstown, New York, that he had been the first black coach in the major leagues. "In the film he appears repeatedly to comment on the sport and the racism that long marred it. Burns calls him the film's 'conscience.'"[62] A case in point is his commentary in episode seven on the demise of the Negro League: "I welcomed the change because this is what I was thinking about since I was that high. Rube Foster was thinking about this before I was born. The change would make it the *American pastime*" (his emphasis). Later on he adds, "I think we are the cause of the changes. Some of the changes that've been made were because of us. We did our duty. We did the groundwork for the Jackie Robinsons, the Willie Mayses, and the guys that are playing now. So why feel sorry for me? We did our part in our generation, and we turned it over to another generation and it's still changing—which is the way it should be."

In areas other than race, though, several journalists reacted viscerally to the higher purpose that Burns seemed to be forcing onto *Baseball* in the many media interviews he was providing nearly nonstop around the country, such as referring to "this game [as] the bellwether of the republic" at a September 12 National Press Club appearance.[63] His rationale was that "there was, I felt, a rather snobbish suspicion of the historical legitimacy of a series on baseball. Many people thought it was frivolous and not as relevatory as we believed, and therefore I felt it required a redoubling of my effort to convince. Sports histories have not reached the mainstreams. The metaphor that I was choosing to tell about baseball needed to be sold."[64] In the end, however, there was an evident backlash against Burns's hard sell which affected the critical reception to the miniseries where "even the most enthusiastic preshowing reviews of *Baseball* observed that it was a tad overdone."[65]

Popular criticism of *Baseball* tended to center on Burns as a producer-director as well as his aspirations for the series as a whole. A characteristic example appeared in *Commonweal:*

> The fellow has a D. W. Griffith complex. Griffith, the greatest of silent directors, transformed forever the art of the movies, but was plagued by a Protestant/universalist compulsion to turn whatever he was filming into a Big Statement on Everything: these cosmic hungers are part, of course, of being gifted at all . . . [s]o overweening is [the program]'s moral urgency, though, to make baseball a template for the social history of the American century that Burns's documentary becomes—I don't know another way to say this—spiritually muscle bound.[66]

An enthusiastic if exasperated *Newsweek* critic likewise wrote that "baseball can't be something so frivolous as a game. It's nothing less than the American 'Odyssey' . . . the Rosetta stone . . . a Blakean grain of sand that *reveals the universe* [his emphasis]. Whew. At its most ponderously pretentious, 'Baseball' offers the best reason yet for banning intellectuals from the ballpark." This reviewer, nevertheless, concludes, "when this 18½-hour documentary doesn't overreach, when it goes with what comes naturally, it delivers plenty of strikes . . . the Kid should be proud. If 'Baseball' isn't 'The Civil War'—and what is?—it certainly gives us everything we ever wanted to know. About a *game* [his emphasis]."[67]

The cumulative audience for *Baseball* was the second highest in public television history after *The Civil War,* averaging a 5.5 rating over nine evenings of programming as compared to the PBS prime-time average of 2.2.[68] Mixed with the social commentary, *Baseball* also includes many scenes of vicarious excitement and personal charm. Two examples run back-to-back midway through the "Seventh Inning." The first, entitled "The Catch," is a 12-minute biographical sketch of Willie Mays, similar to the other career vignettes that appeared earlier in the series on such seminal ballplayers as Ty Cobb, Honus Wagner, Cy Young, Walter Johnson, Babe Ruth, Shoeless Joe Jackson, Lou Gehrig, Ted Williams, Josh Gibson, Joe DiMaggio, and, of course, Jackie Robinson. The grace and enthusiasm of Mays is on full display in a spirited montage, edited to a swing tune dedicated to the "Say Hey Kid," showing his prodigious talents in every aspect of the game, including hitting with tremendous power, running the bases with speed and style, and throwing out opposing runners from deep in center field. "To many," John Chancellor affirms in voice-over, "he was the greatest player who ever lived."

The scene climaxes with Willie Mays's famous over-the-shoulder running catch in the 1954 World Series, one of the truly legendary feats in baseball history, on a par with Ruth supposedly "calling his shot" in the 1932 World Series and Bobby Thompson's "shot heard 'round the world" to win the National League pennant in 1951. The catch occurred on September 29 with the score tied 2-2 in the eighth inning of the first game in a series which the Cleveland Indians were heavily favored. Vic Wertz lifted a towering fly ball to deep center field which looked unreachable to many in the ballpark and on television. Mays not only caught the ball at full speed but in one motion turned like a corkscrew and threw it back to the infield so the runners couldn't tag up and score. "Something seemed to go out of the Indians after Mays's extraordinary catch," concludes Chancellor, "The [New York] Giants went on to win the opener and then the series in four straight games."

Robert W. Creamer, one of the onscreen commentators and advisors on *Baseball*, remembers, "I'd say to [Burns], 'Well, that story is myth.' Ken would say, 'sometimes myth is as important as historical fact.'"[69] In the case of Mays's catch, broadcaster Bob Costas provides the more romantic account. He calls the running grab an "optical illusion," saying "it was more than just a great acrobatic play. It was a play that until that point was outside the realm of possibility in baseball." Cleveland Hall-of-Famer Bob Feller, clearly not an impartial witness either, offers a more skeptical and subdued rendering: "That was far from the best catch I've ever seen. It was a very good catch but we knew Willie had the ball all along . . . Willie is a great actor, a great ballplayer." Throughout the series then, Ken Burns, the producer-director as popular historian, regularly presents both the myth and reality of *Baseball*'s legendary figures and incidents, leaving it up to the audience to decide which version of events it wants to believe.

The following 2-minute 50-second scene, "7th Inning Stretch," similarly employs an effective viewer involvement strategy. In this brief interlude built around *Baseball*'s anthem, "Take Me Out to the Ball Game," eleven of the series' experts and witnesses let down their hair before the camera. In a scene that generates the awkward charm and silliness of a family get-together, the performers reveal themselves with warmth and good humor: Harvard paleontologist Stephen Jay Gould stops and starts at the beginning trying to hit the right key; next, Billy Crystal, does a mangled imitation of Mickey Mantle singing; poet Donald Hall comes to a complete stop when he forgets the words;

two former Negro-Leaguers, each crooning different lyrics, laugh with delight; Bob Costas, Robert Creamer, and Curt Flood sing a couple of words each, as Roger Angell finishes the verse with "I can't carry a tune"; writer Daniel Okrent and historian Doris Kearns Goodwin continue, as she blushes, "this is pretty embarrassing"; and Buck O'Neil concludes by saying "I'm not much of a singer" before stealing the show with a soft, heartfelt rendition of the song, bringing it all back home. The scene ends with a vintage newsreel version of "Take Me Out to the Ball Game" where the announcer invites the viewing audience to "follow the bouncing ball and sing."

The strength of *Baseball*, in this way, rests mainly in the individualized success of specific scenes—some historically informative, others mythically resonant, with a few even designed to just touch and entertain the audience—rather than in how the 18-hour 30-minute storyline holds together in its entirety. The 11-hour 30-minute *The Civil War* in contrast marches slowly but inexorably toward its inevitable conclusion with a final Union victory and Lincoln's assassination. *Baseball* has no logical resolution that it's heading toward. In the prologue to inning nine, for instance, John Chancellor announces, "everything had changed, and nothing much had changed," suggesting the cyclical nature of the miniseries. The one exception is the racial subplot which starts with the exclusion of blacks in inning one, crosscuts between the white major leagues and the old Negro League in inning five, climaxes with the triumphant career of Jackie Robinson in innings six and seven, and then is brought to a fitting conclusion with Robinson's death and funeral in inning nine. None of the other subplots in *Baseball*—for example, the fortunes of the Red Sox or the Dodgers, or the chronic struggle between labor and management, or even the legendary exploits of the game's greatest stars—possesses a similar kind of closure: thus, they are inherently episodic in nature.

The open-ended structure of the "Ninth Inning: Home" with its moving requiem for Jackie Robinson unintentionally transforms the last episode into a film elegy for baseball's fading importance. In the short term, at least, the sport has never recovered from the calamitous 1994 strike, despite the brief upswing in interest during Mark McGwire's and Sammy Sosa's 1998 home-run chase in which they surpassed Babe Ruth's single-season record of 60. "As baseball fades," according to columnist and long-time fan Charles Krauthammer, "it turns to memory to keep itself alive. The game's main refuge is its glorious receding past, to which it repairs with ever increasing frequency and desperation."[70] Ken Burns's *Baseball* skillfully captures this illustrious history. The game was the national pastime during the American century, central to the fabric of society and culture and, in retrospect, much more than its present position as just one of three major sports in the United States competing against each other for fans and highly coveted television contracts.

AN EXTENDED FAMILY APPROACH TO FILMMAKING

The best part of making the films the way we do is that we have been able to sort of home grow a lot of people who came in and work . . . and have now gone on to do their own work that is hugely satisfying.

—*Ken Burns, 1999*[71]

Ken Burns was directing the final day of audio mixing on *Baseball* at Sound One in the Brill Building, just off of Times Square in New York City, in late April 1994. As the session wound down, the creative team was simultaneously exhilarated and exhausted, having finally brought the production phase of their mammoth 18-hour 30-minute miniseries to completion. Burns opened several bottles of champagne to celebrate and "there [wa]s a full hour of toasting and tears," as he lifted a glass and glanced around the room at many of his closest associates during the last three years, declaring "'I love everyone in the room— my family.'"[72] "When you start as a filmmaker," reminisced mixer and long-time collaborator Lee Dichter, "you try to do everything yourself, and then through the years Ken has gotten better at utilizing the people around him, giving them the freedom and also understanding how much they can contribute—the whole group of people."[73]

Burns has not been without his critics over the years. When *The Civil War* was first released, for example, *Newsweek* quoted an unnamed historical advisor who had worked with the producer-director as saying "the problem with Ken is he can't stand to share credit."[74] A flurry of rumors similarly surfaced when "Ric Burns went out on his own after what some say was a falling-out over credit for *The Civil War*," forming his own New York–based company, Steeplechase Films in 1989.[75] By and large, though, he has "inspire[d] a respect bordering on adulation" among his immediate staff at Florentine Films/American Documentaries.[76] Supervising editor Paul Barnes, for instance, reports "I ha[ve] never worked with a film director that I [am] so simpatico with. We just ha[ve] this amazing symbiotic working relationship which has continued for all these years."[77] Burns, too, now makes a point every chance he gets of acknowledging "my art is a collaborative one. What you see standing before you is really the tip of the iceberg . . . There are many, many people who own equal credit to a film by Ken Burns."[78]

Baseball coproducer Lynn Novick also admits that there are occasional "disagreements and arguments, but they're part of the creative process. Ken's single-minded, but he brings out the best in people. I've worked with a lot of producers and I've never known one who let everyone get their two cents in like he does. He's open because he's confident."[79] Burns likewise describes the intense but intimate and caring working conditions that he and his colleagues have built for themselves over time: "I have worked with the same editor, Paul Barnes, for more than a dozen years. I've worked with the same cameraman for more than twenty years, Buddy Squires. I've worked with coproducers going on a decade. Essentially this is not a business but a family. We have and enjoy the benefits of a community that I wouldn't have any other way."[80]

Burns turned increasingly toward his stock company of associates to share both the opportunities and responsibilities that emerged after the successes of *The Civil War* and *Baseball*. The idea for *The West* first emerged in a series of ongoing conversations between the filmmaker and James Dougherty, director of the Public Programs Division at NEH, during the fall of 1985. At the time, they were discussing Burns's funding proposal and developing plans for *The Civil War* when "they agreed that, with the possible exception of [that conflict], no other subject could provide such a revealing look at the essence of the American experience" than "a history of the West."[81] *The West* remained on the back burner at Florentine Films/American Documentaries for four more years, until Burns approached Time-Life Video about distributing *The Civil War* in July of 1989. While producing this documentary, Burns and his colleagues had often referenced Time-Life's 26-volume educational book series on the history of the Civil War which provided "hundreds of first-person quotes and thousands of photographs for the filmmakers to consider."[82] Time-Life had also published a 26-volume companion series on the West and thus immediately "indicated interest in acting as a producer in such a [follow-up film] venture," especially after Burns relayed to company officials that Jim Dougherty of NEH was similarly enthusiastic about sponsoring part of this programming.[83]

Over the next 14 months, Ken Burns was, nevertheless, preoccupied with the promotion and release of *The Civil War*, the production of *Empire of the Air*, and the preparation of *Baseball*. In October 1990, he "and Time-Life picked up the pace of their negotiations" following the "spectacular success" of *The Civil War*.[84] Geoff Ward once again agreed to serve as head writer, and both he and Burns interviewed several candidates over the course of a month who would step in to produce and direct the series. The unanimous choice, in the end, was Stephen Ives. As Burns recalls,

I had the idea, the impossible desire to make a film on the history of the West after *The Civil War,* but I was already committed to *Baseball* and enlisted Geoff . . . we needed to find someone who was capable of doing the day-to-day, the producing and directing, and that was Steve, who I'd worked with on *The Civil War* briefly and *Baseball* briefly—and he and I had produced a history of the Congress and he made a good film on Lindbergh, which Geoff wrote. So we've been sort of tied to each other intricately for almost a dozen years and *The West* is the latest fruit of that collaboration.[85]

After coproducing *The Congress,* Stephen Ives assumed primary responsibility for *Lindbergh: The Shocking, Turbulent Life of America's Lone Eagle,* which premiered as the opening episode of *The American Experience*'s third season in 1990. Ives had grown up in Lincoln, Massachusetts, the son of David O. Ives, former president of PBS affiliate WGBH in Boston. Ives had studied history at Harvard but did not even consider filmmaking until after graduation when he moved to Texas "wanting to break away from the East."[86] "That's when I became fascinated with the myth and landscape of the people of the West," he recalls.[87] He, in turn, pursued his new-found regional interest by producing, directing, and co-authoring his first film project, *Where the Heart Is* (1986), which explored the meaning of home and family to Texans. This initial effort won recognition at the Chicago International Film Festival but, more importantly, led to his introduction and association with David McCullough.

While making *Where the Heart Is,* Ives wrote to McCullough who was then the host of *Smithsonian World* (PBS, 1983-1988), inviting him to narrate his new film. After some negotiation, the author accepted, and Ives eventually met Burns through his working relationship with McCullough. At the time, Ken Burns was starting production on *The Civil War* and preparing *Thomas Hart Benton* and *The Congress.* He hired Ives to serve in various research and filmmaking capacities on all three projects. "It was my baptism into the world of historical documentaries," recounts Ives, "and it was a hell of an awakening. Ken makes films with a tremendous intensity, and at the time he was making *three.* It was a crazy, chaotic time, and I got thrown right into it."[88] Ives contributed well enough to earn an associate producer credit on *The Civil War* and to coproduce *The Congress* with Burns.

Once *The Congress* was released and the production phase of *The Civil War* was nearing completion, Ives suggested that he produce and direct a film biography of the celebrated American hero Charles A. Lindbergh under the auspices of Insignia Films, a small independent production company that he had founded in 1987. McCullough provided a personal entrée to the aviator's widow,

Anne Morrow Lindbergh, having been the first person to interview her on-camera about her life for a 1984 Emmy award-winning episode of *Smithsonian World*. Burns also provided his full backing and endorsement to Judy Crichton, the executive producer of *The American Experience* at WGBH. Burns, moreover, agreed to serve in a supervisory capacity as coproducer. "Ken was a mentor to this project and to me," professes Ives, "I got into historical documentary films because of Ken."[89]

Lindbergh was well received by audiences and critics, reflecting many of the essential aspects of Burns's now well-recognizable style, including Geoff Ward's literate script read by narrator Stacy Keach, a tight, lean 54-minute biographical storyline, the rephotographing of contemporary imagery, and the incorporation of period music, among other familiar elements. The film won a CINE Golden Eagle and a Red Ribbon at the American Film Festival; Walter Goodman of the *New York Times* declared it a "sensitively made documentary that . . . captures the public and private Lindbergh"; and Martin Zimmerman of the *Los Angeles Times* noted that it was "a powerful slice of history . . . an engrossing study of a complex figure."[90] Burns felt, as a result, that "[n]o one could be more prepared for *The West*, and the telling of that complex story, than Ives."[91] Stephen Ives, for his part, was exceedingly comfortable that "Ken and I had tried out this creative partnership before, successfully, in producing a film on Charles Lindbergh for PBS's *The American Experience* series, which is why we were confident that we could take on something 10 times as big as *Lindbergh*."[92] By December 1991, Ives had started researching and closely collaborating with Geoff Ward on the first draft of *The West's* multipart script, as Burns turned his attention to promoting the upcoming telecast of *Empire of the Air* and immersing himself ever deeper in the even larger miniseries *Baseball* with its fall 1994 debut date.

Unbeknownst to anyone at Florentine Films/American Documentaries, Ric Burns was also planning a documentary history of the West. When both brothers learned that they had separate projects in the works during the late 1991 holiday season, "we were sort of surprised and startled," confesses Ric, "there was some head-scratching, and we wondered for a while if it wouldn't be best if we joined forces."[93] As Ken and Ric reflected, however, they realized that many of the key above-the-line people were already in place on both films, and the respective fundraising efforts were also well underway. The lateness of their joint discovery, furthermore, precluded any discussion of the thorny issue of who would ultimately be in charge. "We worked together five years," explains Ric, "If you wanted to go to the best university and learn how to make a film, go and make a film with my brother. It was the greatest experience in the world, but having gone through it, you would not want to not control the film."[94]

After Ken Burns's production crew wrapped *The Civil War* in 1989, Ric, who was relatively unknown at the time, visited Judy Crichton with a proposal to make a historical documentary on Coney Island for *The American Experience*. His chief motivation was to assume more authorial supervision than ever before, developing his own projects as the main producer-director. For his first time out in this regard, Ric planned to adapt John F. Kasson's *Amusing the Millions: Coney Island at the Turn of the Century* (1978), a concise and richly illustrated analysis which ably revealed this celebrated American landmark and amusement park as a harbinger of the country's new emerging mass culture.[95] He was also teaming with two of his Florentine contacts, Buddy Squires as coproducer/cinematographer and Paul Barnes as editor, along with Richard Snow, then managing editor of *American Heritage* magazine who would serve as the scriptwriter. The finished film, *Coney Island* (1991), is an absorbing 72-minute portrait of the rise and fall of the park; it immediately established Ric Burns as an important up-and-coming documentarian in his own right, winning for him and Squires the prestigious Erik Barnouw Award in 1992 from the Organization of American Historians. His next production would further solidify his growing reputation while also clarifying the subtle differences between his own creative outlook as a producer-director and that of his brother, Ken.

Ric Burns named his company Steeplechase Films, after the first of three successive amusement parks constructed at Coney Island, thus affirming for himself the importance of this first self-initiated production in his firm's title. His principal partner since its formation has been his coproducer Lisa Ades, who worked previously at Thirteen/WNET on local New York public affairs series, such as *Metroline* and *The Eleventh Hour*. She quickly demonstrated her considerable producing abilities on *Coney Island* and, consequently, became an essential collaborator with Ric on his developing plans for a much more extended and ambitious series on the West. While conducting preliminary research, Burns and Ades came across a notorious and harrowing story from America's past which instantly caught Ric's attention. He was intensely drawn to the tale of the ill-fated Donner Party, the group of 87 men, women, and children who set out on a 2,500-mile expedition from Springfield, Illinois, headed for Sutter's Fort, California. The journey gradually regressed into a grim struggle for survival, involving death and cannibalism, leaving only 46 weary and emaciated emigrants alive at the end of their year-long ordeal. The broader significance of the Donner narrative immediately attracted Burns, as he describes:

Here was an event that took place within the memory of our grandmother's grandmother. I mean measured by any realistic yardstick, an event that took place very, very recently. And it was a time [1846] when the country was still so nascent, so underformed, so much in the process of becoming what it was going to become . . . here sort of like a *dark shining parable* at the very origin of California of the Western dream is this tale of people who are reaching too far too fast, not because they're bad, but because they're human. And I think the story tells us a lot about the possibilities but also the perils of what we now call, glibly, "The American Dream."[96]

Burns, in fact, was so compelled by the account of the Donner Party that he pitched the idea to Judy Crichton for possible inclusion in *The American Experience*. He did not want to restrict his telling of what for him was an emblematic moment in Western history to an abbreviated 10 to 15 minute sequence lost within a much larger epic plot line for his upcoming miniseries on the West. Crichton admits her first reaction to his proposal was highly skeptical: "I was terrified of the subject matter and I didn't know how the hell he was going to pull it off. Margaret Drain [then senior producer for the series] just leapt on faith, and the two of them talked me into it."[97] Welcoming the challenge, Ric Burns's storytelling skills reached new heights in his reconstruction of this legendary American tragedy. Having only 20 available period images to work with, he and Ades expanded their formal repertoire out of necessity, moving well beyond what had now become the PBS house style for historical documentaries established by Ken Burns. They successfully imbued *The Donner Party* (1992) with an almost hypnotic, dreamlike quality by utilizing an assortment of expressionistic techniques such as mood-enhancing filters, slow-motion cinematography, and a haunting electronic score. *The Donner Party*, as a result, was an even greater critical success than *Coney Island*, garnering a George Foster Peabody Award for excellence in programming and two Emmy nominations for Ric's writing and direction. The younger Burns had not only established his own filmmaking credentials in just four short years, he was now one of a handful of featured producers at *The American Experience*.

Hollywood filmmakers had actually rekindled a small revival in Western-influenced stories and themes beginning in the mid-1980s into the 1990s with a series of films profiling Native American characters and concerns, such as *The Emerald Forest* (1985), *The Mission* (1986), *Dances with Wolves* (1990), *Black Robe* (1991), *At Play in the Fields of the Lord* (1991), *Thunderheart* (1992), *The Last of His Tribe* (1992), *Legend of Wolf Mountain* (1992), and *The Last of the Mohicans* (1992). Emerging from this entertainment context, baby boomer filmmakers were now attempting to come to grips with the colonial past of the Americas and the diversity of races and nationalities that are part of that history.

A new generation of scholars was likewise sharing this renewed preoccupation with western history and myth, revising it from an entirely fresh perspective and sensibility. As historian Alan Brinkley explains,

> Frontier history is the story of the contact of cultures, their competition and their continuing relations. It cannot be the story of any one side. "Multiculturalism," which is becoming the source of sweeping (and often painful) reappraisals of almost every area of American life, has a special claim to attention for historians of the West. No other region has had so long and intensive an experience of racial and ethnic diversity; no other place displays the imprint of multiculturalism more clearly.[98]

The older conception of the classical Western as a success story celebrating white America's creation of its transnational civilization was conspicuously absent from this new crop of films. According to film scholar Thomas Schatz, this genre was once the country's *"foundation ritual* endlessly repeating the great triumph of settlers against the hostile wilderness . . . the Western project[ed] a formalized vision of the nation's infinite possibilities and limitless vistas, thus serving to 'naturalize' the policies of westward expansion and Manifest Destiny."[99] The widespread popularity of this mythic structure cannot be underestimated. "During its high point as a film genre from around 1940-1960, as much as one fourth of the annual Hollywood output consisted of Westerns."[100] In the late 1950s, 30 Western series dominated prime-time television, comprising 7 of the top-10 shows. It is not surprising, then, that a new cohort of filmmakers and academics in their thirties and forties were now deciding to revisit this extraordinarily popular story form from their adolescent and teen years.

According to cultural critic John Cawelti in the second edition of his seminal analysis *The Six-Gun Mystique,* the decline of the Western after 1960 was the result of three increasingly anachronistic aspects of the genre. First, the orderly ritual of violent action centered around the gunfight appeared "sanitized and pale" against the "randomness and ambiguity" of contemporary expressions of violence onscreen.[101] Second, the deep-seated sexism in the traditional Western was alienating to at least half the potential audience as a result of sweeping attitudinal changes generated by the women's movement in the 1960s and 1970s.[102] Most importantly, though, the genre's unsympathetic and racist presentations of Native Americans, Latinos, and blacks rendered the story form irrelevant and untenable as civil rights progress in society as a whole left the Western far out of step with most Americans.[103] In contrast, the 1990s resurgence of big- and small-screen Westerns was concerned with redressing racial and ethnic stereotypes first and foremost, although corrections in gender representations do occasionally seep to the surface as well.

Ric Burns's *The Way West* is a prime example of this new revisionist model. His documentary categorically rejects the "foundation ritual" that celebrated nation building by portraying the Indians as aggressors and the whites as victims, thus inverting the historical reality and justifying the ultimate conquest of Native Americans. From this reform perspective, native peoples and nonwhites especially are presented as being displaced from their homes and subjugated in the wake of the relentless move west of Anglo-American emigrants. Although hints of this alternative narrative can be traced as far back as the mid-1960s in Hollywood films such as John Ford's *Cheyenne Autumn* (1964), its ascendancy as the standard model was never fully confirmed until Kevin Costner's *Dances with Wolves* won the Best Picture Oscar in 1991 and grossed $250 million at box offices worldwide.

Costner also hosted an eight-part CBS documentary series entitled *500 Nations* (1995), chronicling the history of indigenous peoples in North and Central America from pre-Columbian times to the defeat of the Plains Indians at the end of the nineteenth century. This nonfiction special along with similar ones produced by Ted Turner's WTBS and The Discovery Channel all predated Ric Burns's four-episode *The Way West* by at least six months during the 1994-1995 TV season. The overall reception to this programming trend was generally mixed, as one *New York Times* journalist , who was quoted in *Newsweek,* "attacked the recent spate of pro-Native-American documentaries as 'affirmative action TV,'" while *Time*'s television critic surmised that "if the Old West is back, it's not necessarily the West of old. Call it political correctness or a long-overdue historical corrective, but [today's] picture of the West has a grubbier, less celebratory, more multicultural look this time around."[104] Burns's prime-time special presentation for *The American Experience,* which premiered on two evenings in May 1995, fit comfortably within this newly revised concept of the TV Western.

Burns and his creative team originally intended *The Way West* to be three hours long, but the $1 million production eventually doubled in size because of the sheer amount of research material that they uncovered, thus delaying its debut by nearly a year. The miniseries, in essence, is a solemn, heartbreaking lament for the centuries-old Native American way of life that was irrevocably shattered by whites on the way west. The story basically traces the encounter of irreconcilable civilizations in which the outcome of this clash of cultures is never in doubt. The first two 90-minute episodes, telecast on May 8, cover the early westward expansion from 1845 until the completion of the Transcontinental Railroad in 1869. "There was something about the railroad," reveals writer and onscreen commentator Tom McGuane, "that gave Native Americans a sense of how dire the end was that awaited them."

The last two episodes, also 90 minutes each, largely follow the Indian wars on the Great Plains from 1870 through the battle of Little Bighorn in 1876, culminating with the 1890 massacre of an estimated 250 Lakota men, women, and children at Wounded Knee. The conflict between the soldiers and settlers from the East and the native peoples of the West was tragically one-sided. There were 10 million inhabitants of the continent "when whites first arrived," reports narrator Russell Baker, "300 years later, 90 percent of the population was wiped out." The final sequence of the miniseries ends in 1893 with a special emphasis placed on historian Frederick Jackson Turner's famous pronouncement that the western frontier was now officially closed forever. Although conventional in its epic structure and the time frame it covers, *The Way West* broke new ground in respect to the old classical model by resolutely presenting its story from the point of view of the indigenous inhabitants of the West rather than from the position of the white newcomers.

In reviewing the film for *Newsweek,* Rick Marin perceptively characterized *The Way West* as "a masterly piece of nonfiction, less a documentary than a tragedy in four 90-minute acts."[105] As writer, director, and coproducer, Ric Burns concisely encapsulates the entire plot line in just five sentences at the beginning of episode one entitled "Westward the Course of Civilization Takes Its Way." Only 3 minutes and 20 seconds into the 6-hour miniseries, Russell Baker summarizes the forthcoming story in a brief voice-over while a curious and disturbing photograph of a lone white man draped in an American flag, pointing his pistol off into the distance, appears onscreen:

Americans had been dreaming of the West for centuries, when in 1845 a newspaper editor in New York uttered a fateful declaration. "It is our Manifest Destiny," he wrote, "to overspread the continent that Providence has allotted for our yearly multiplying millions." One year later the nation exploded westward to fulfill that destiny, and the most troubling and transformative episode in American history had begun. Between 1845 and 1893, the American West was lost and won. The vast continent Thomas Jefferson was sure would take a thousand years to settle and subdue was wound with ribbons of iron and wire and brought within the dominion of the United States. Along the way, the lives of hundreds of thousands of Native Americans were violently disrupted and all but destroyed.

This opening scene succinctly infuses a portentous sense of impending catastrophe which aptly typifies the rest of *The Way West*. Ric Burns's clear-cut narrative focus signals both the great dramatic strength and the conceptual limitation of his documentary.

The opening scene to the fourth episode, entitled "Ghost Dance," is another case in point. This 3-minute 15-second section begins with a fast motion shot of rolling clouds which quickly dissolves into a majestic view of the pink western sky filled in slow motion with flying birds. Actor Wes Studi's voice fades up reading the words of Lakota holy man Black Elk: "While I stood there I saw more than I could tell, and I understood more than I saw, for I was seeing in a sacred manner the shape of all things as they must be, living together in one being, and I saw that the sacred hoop of my people was one of many hoops that made one circle as wide as daylight and starlight, and in the center grew one mighty flowering tree, and I saw that it was holy [cut to a medium shot of Black Elk]."

Brian Keane's exquisitely sad and moving theme song for the series then fades up over a sepia still image of a western town with literally hundreds of people milling about the main street. Russell Baker, once again, ably sets the scene:

> The West was changing. For nearly four centuries, European-Americans had pursued their destiny across the wilds of the North American continent. Now what was wild about the West was beginning to recede. [Cut to a photograph of railroad tracks running alongside a high peak.] Hundreds of visible and invisible lines now cut across what was once unbroken space [cut to a first-person camera shot taken from the front of a train making its way through the mountains]—state lines, telegraph lines, railroad tracks, roads, fences. By 1876, it was clear to most Americans that the vast spaces once seemingly infinite, were closing in [moving camera shot ends as the train enters a dark tunnel].

The scene climaxes with the camera point of view slowly tilting upward to reveal a solitary white man standing alone on a mountaintop. The voice of George Plimpton fades in as that of poet Walt Whitman: "I look off the shores of my Western sea, having arrived at last where I am, the circle almost circled, but where is what I started from so long ago? And why is it yet unfounded?"

The lyrical beauty and effect of this scene is undeniable. Ric Burns won an Outstanding Achievement Award from the Writers Guild of America for his poetic script which, in this instance, elegantly and economically grounds the viewer in a West of shrinking dimensions and opportunities, thus making the eventual contrast between Black Elk and Walt Whitman that much more striking and effective. The scene is basically structured in three parts, starting with the inspirational words of Black Elk. This is the most visually arresting section, shot mainly in soft impressionistic colors and given an otherwordly cast with its mixture of fast and slow moving imagery. The Lakota mystic is one of the wisest spiritual barometers in the series, which is telling since he is standing on the losing side of Manifest Destiny and of the war for the Black Hills.

The next two-thirds of the scene ironically underscore the discontent of the winning side. The middle portion, featuring Russell Baker's narration, suggests a West that was changing for the worse. The huddled image of the townsfolk imply an overcrowded condition that is at odds with the pristine wide-open spaces of the natural environment. The West is now crowded with people and arbitrary boundaries—"state lines, telegraph lines, railroad tracks, roads, fences." Much more has evidently been lost by whites on the way west than won. Even Walt Whitman, Anglo-America's prototypical poet-seeker from the second half of the nineteenth century, appears alone and adrift on his increasingly melancholy quest for greater awareness and contentment. Ric Burns's scenario, in the end, is not wholly untrue; *The Way West* does indeed assert a point of view long suppressed for much of the twentieth century. The single-minded debunking of the classical Western formula, however, characteristic of this miniseries as well as of the wider cycle of revisionist films it belongs to, merely replaces the old foundation ritual with a darker counternarrative, which foregoes in its linear construction a subtler and more complicated rendering of the historical record.

In this scene specifically, Black Elk, who fought the whites on the Great Plains and survived Wounded Knee, has a vision of reconciliation. In the next century, he would achieve near guru-like status among members of the Native American Church, radical environmentalists, and various new age groups following the publication and wide dissemination of John G. Neihardt's *Black Elk Speaks: Being the Life Story of a Holy Man of the Ogalala Sioux.*[106] Even though Black Elk's undeniable mysticism has come to embody the more traditional expressions of Native American religious practice, which clearly was stifled and censored by white missionary control, he was, in fact, a Roman Catholic convert himself and a catechist on his reservation. The words of Walt Whitman, too, are taken from a drastically altered version of his 1867 poem "Facing West from California Shores," which has almost nothing to do with the American West at all but is concerned with the spiritual quest of the wandering poet who experiences both the joy of growing awareness and the simultaneous emptiness of realizing that complete fulfillment is probably always beyond his grasp. The still image of the lone man standing on the mountain peak, moreover, is of photographer William Henry Jackson surveying the view from Glacier Point, Yosemite Valley sometime during the 1880s, thus totally refashioning the essential content of both the poem and picture.

Poetic license is, admittedly, unavoidable in historical filmmaking and television production; such practice goes with the territory. Still, the reasons why producers and directors as popular historians take liberties with their filmic raw materials are matters of the utmost importance. Too often in *The Way West*, words and images from the past are reshaped and channeled into a largely unidirectional plot line, which successfully propels the action forward with an ever increasing momentum but fails to show the many sides and faces of one of the most complex and pluralistic chapters in American history. Ric Burns acknowledges as well that he "is drawn to catastrophes a great deal more than Ken is [who] is not necessarily [attracted to] the upside of life [as much as] stories in which adversity is overcome in the long run . . . and for me," Ric continues, "I don't know what it is, some kind of instinct for disasters [that] I find myself compelled by in the subjects I'm drawn to."[107]

Even after the release of *Baseball*, Ken Burns was managing many projects simultaneously. He was guiding *The West* to completion as its executive producer, developing the *American Lives* series of historical profiles, and launching a huge history of jazz. Burns's solution was "to give some of this away," sharing his production ideas and opportunities with his long-time colleagues at Florentine Films/American Documentaries.[108] "I was interested in sharing more," concedes Burns, "this seemed to be the perfect chance."[109] "Ken's probably evolved in that respect," adds his brother Ric, "I'm not sure that I have yet, though. When you have the driver's wheel, it's really tough to let it go."[110]

Ric, who is regularly asked about his older brother in many interviews, was questioned often about Ken's work on *The West* since Ric had just finished his miniseries on the same topic. Ric's standard response was that Ken "will have no more or less to do with *The West* than an executive producer ordinarily would, [although I] understand . . . why Ken's name is so often attached to the project, but *The West* will sink or swim because Steve Ives is a great director, not because Ken is an executive producer."[111] In fact, Ken Burns and Stephen Ives were working out their respective roles on *The West* as they went along since the project was so much larger and more complicated than *Lindbergh*. As their project neared the final phases of postproduction in the spring of 1996, producer-director Ives noted that "it's been an interesting evolution for both of us."[112] Burns also confessed, "I was initially worried that my relationship to this series wouldn't be one I could give my whole heart to. I'd never been a foster parent on a project before, I'd always been the biological parent, the author." In his new role as executive producer and senior creative consultant, Ken Burns was "in-the-trenches, [with] hands-on experience" during the scripting process; and then again, he played a significant part in the editing and postproduction phase, but "I didn't have anything to do with how the scenes were built. It has Steve's look; what I did was guide it."[113]

The West was originally planned as a seven-part, ten-hour program. After 20 drafts of the script by Geoff Ward who was joined by Dayton Duncan as co-writer in the summer of 1991, they added an eighth episode, making the series 12 hours 30-minutes. Stephen Ives's creative team also had, as usual, the advice of a panel of academic consultants "whose eye for historical detail kept our stories clear and our chronology on track," reports producer-director Ives.[114] These scholars were particularly influential in making sure The West went beyond "whites invaded the continent and took it away from the Indians," remembers historian Richard White.[115] Many of these advisors made recommendations similar to those of Ramón Gutiérrez, then professor and chair of ethnic studies at the University of California, San Diego, who, after reading an early draft of the script, wrote "while I am happier with the treatment of Blacks, Asians and Hispanics in this version too, there is still far from an adequate coverage. There is still a cowboy/Indian bias to the program. You give a great deal of attention to Kit Carson, Fremont, but never is there a sustained treatment across the program to a person of Chinese, Mexican or black ancestry. Unless you do something about this, you will simply be reiterating the old colonial assumptions about the history of the area."[116]

The filmmakers, in response, widened their perspective and their time frame considerably. The West is far more expansive, in this regard, than the wave of revisionist fictional and nonfictional Westerns that preceded it. The sweep of its narrative begins in 1528 with the landing in Texas of the first conquistador, Cabeza de Vaca, and ends after the great Los Angeles water swindle in 1913. In between, the filmmakers integrate both epic and biographical strategies with the intention that the smaller personal stories will ultimately evoke and further clarify the much larger sociopolitical and cultural currents. Seventy-four different individuals are accorded their own vignettes throughout the course of The West, while more than a third of these historical figures are people of color or women. Ward and Duncan's script establishes this precedent at the outset with narrator Peter Coyote outlining the multicultural context only 3 minutes and 50 seconds into Episode One (to 1806) entitled "The People": "People came from every point of the compass. To the Spanish who traveled up from Mexico, it was the North. British and French explorers arrived by coming South. The

Chinese and Russians by going East. It was the Americans, the last to arrive, who named it the West. But to the people who were already there, it was home—the center of the universe."

Ives, Burns, Ward, and Duncan, moreover, made a concerted effort to initiate a synthesis between the foundation ritual of the past and the newer revisionist version of white conquest and dominance over the Indians, which was now the operative narrative model for the current generation. Early in their research, the filmmakers latched onto two metaphors that helped them move beyond these opposing stereotypes. The first came from novelist and popular historian Wallace Stegner, who had served as an advisor to *The West* film project until his death in an automobile accident in 1993. Stegner formulated the expression "geography of hope" to allude to the various quests for freedom, wealth, a new start, and other American dreams and desires that emigrants projected onto the landscape as they ventured west. The second metaphor was devised by one of Stegner's former creative writing students, the renowned Kiowa poet and novelist N. Scott Momaday, who also emerges as a central onscreen commentator for the series. Momaday portrayed the annihilation of the buffalo, which plummeted in numbers over the course of the nineteenth century from approximately 65 million to just a few hundred, as a "wound in the heart." "It was a wonderful way to describe the price paid for settlement of the West," explains Ives, "the tension between the 'geography of hope' and the 'wound in the heart,' is a dichotomy that has appealed to us as a way of seeing the western experience."[117]

Episode Seven (1877-1887) is named "The Geography of Hope" after Stegner's metaphor, as 4.5 million more newcomers are shown heading west during this decade, driven by visions of a new future, while most of the Plains Indians are concurrently being subdued and relegated to reservations throughout the region. "By 1877," cites Peter Coyote, "there are 40 whites to every one Indian in the West," as subplots abound about an assortment of homesteaders and adventurers ranging from ex-slave Pap Singleton to future President Theodore Roosevelt. Episode Seven ends as it began, attempting to balance the dreams of the region against the reality of conquest as Wild West showman and mythologist Buffalo Bill Cody, whom Richard White characterizes the "one true genius of the nineteenth-century West," is counterposed with an ominous final reference to a dream had by Sitting Bull, who had joined Cody's traveling entourage. A small bird appeared to the Sioux chief and holy man revealing "Your own people, Lakota, will kill you." Sure enough, in the next episode, "One Sky Above Us," the eighth and last installment of the miniseries, covering 1887 to 1914, Sitting Bull is shot dead on December 15, 1890, by two reservation policemen, Bull Head and Red Tomahawk, as tensions were nearing a boiling point over the ghost dancers at Pine Ridge, just two weeks prior to the eventual massacre at Wounded Knee. Richard White once again offers an insightful critique on how to make sense of this lingering paradox between hope and tragedy which permeates *The West* from start to finish:

> The West is about possibilities, and sometimes the price of success of one group
> of possibilities is going to be the ending of possibilities for another group.
> Americans aren't wrong in seeing the West as a land of the future—in seeing
> a land where astonishing things are possible—what they often are wrong about
> is that there is no price to be paid for that, that everybody can succeed or that
> even what succeeds is necessarily the best for everyone concerned. The West is
> much more complicated than that.

Juggling these two longstanding and opposing traditions in both western historiography and mythmaking sometimes proved to be a difficult challenge for Stephen Ives and Ken Burns. "It wasn't always easy," admits Ives, "we had our creative battles."[118] Many of these disagreements had to do with the proper balance between the "geography of hope" and the "wound in the heart." Burns was often on the side of recalibrating the generational tendency to reflexively feature the darker side of the western experience. "You have to tell these things unflinchingly," explains Burns, but "you can't make it unrelenting horrors. What I began to insist on in the editing room is [a] breath" for the audience.[119] Episode Three (1848-1856), "The Speck of the Future," is a prime example of the results of these artistic struggles between Ives and Burns and their combined ability to finally reach common ground by integrating the classical and revisionist perspectives on the West in an entirely new synthesis.

The working title for the third episode was originally "Seeing the Elephant," an "expression from the gold rush that implied both danger and adventure, and indicated that one had suffered an ordeal and confronted the bitter reality often waiting behind the veil of inflated hopes."[120] The gold rush, in this way, becomes the major preoccupation of Episode Three, serving in general as an allegory for the larger western experience. This main storyline is additionally supported by two subplots concerning increased urbanization throughout the region, represented by the growth of San Francisco, and the dire implications of the mass stampede of emigrants across the continent for Native American tribes who resided along the overland trails to the gold fields.

Over a series of static color shots of the big-sky country, narrator Peter Coyote reports offscreen that the land was a dramatically different place before the gold rush: "The West was American in name only. Few people east of the Mississippi were anxious to venture into its forbidding interior. It still seemed too distant, too mysterious, too dangerous." Next a much quicker-paced montage of period photographs introduces a whole host of immigrants and emigrants representing a wide variety of different ethnic groups:

> Then gold was discovered in California and everything changed for the West and the country. Suddenly gold seekers rushed in from every corner of the globe: Chinese peasants pursuing tales of a gold mountain across the ocean; Mexican farmers, and clerks from London; tailors from Eastern Europe and South American aristocrats fallen on hard times. The thin stream of American emigrants crossing the continent became a torrent, thousands upon thousands of optimistic but inexperienced prospectors, willing to leave their homes and families and set out on the long trail to California, hoping to strike it rich and return in glory.

This opening scene slowly eases to its conclusion with an onscreen portrait of William Swain, a peach farmer from western New York state who leaves his wife and family behind temporarily to seek his fortune in the California gold fields. More importantly, he serves as the narrative touchstone of the entire episode, emblematic of the estimated 30,000 Americans who headed west in 1849 alone. According to producer-director Stephen Ives,

> I think making a history of the West is like trying to paint one of those Albert Bierstadt canvases. You need the epic scale. Yet, you begin with small stories—foreground and background, intimate stories that are based on biographies, based on people. Ultimately, if you work with the canvas long enough, the larger picture becomes clear to you and is possible to be rendered. That was the challenge of the film, to find small individual stories that added up to powerful dramatic episodes that had a beginning, middle and end and then to find how those pieces ended up creating one large narrative of the history of the West.[121]

Ives initially discovered the story of William Swain in *The World Rushed In*, written by historian J. S. Holliday, whose commentary figures prominently throughout "The Speck of the Future." The author even provides this episode with its new name in his onscreen reference to the nugget of gold that James Marshall picked up out of Sutter's Creek on the morning of January 24, 1848, officially setting the gold rush into motion.[122] "Jim Holliday is the

Shelby Foote of our third episode," notes Ken Burns, "you think that if we pulled the camera back he wouldn't be in some comfortable office, but that he'd actually be out there with a pick and shovel and that the American river would be rushing by his feet."[123]

One of the differences of opinion between Burns and Ives was over the use of titles to introduce the successive sections of each episode. Ives wanted to forego this well-established element of Burns's storytelling technique as he had earlier in *Lindbergh,* while Burns felt that the inherent complexity of this particular narrative, with its constant demands of shifting back and forth between the "geography of hope" and a "wound in the heart," warranted the insertion of title cards to keep audiences oriented both chronologically and thematically. In the end, they compromised, as Ives employs approximately half the number of title graphics that Burns typically utilizes when he directs.

In Episode Three, for instance, six titles are interspersed throughout the 87-minute 45-second plot structure. Four title graphics are used to frame the story of William Swain (and by implication, all the other emigrant forty-niners he represents), chronicling Swain's overland journey across the country ("My Share of the Rocks"); his many diary entries and the letters he sent to and received from his wife ("Stay at Home"); the inevitable slim pickings in the gold fields ("The Diggings"); and his eventual return home 18 months later by sea from San Francisco with just enough money left to book the passage ("Days of '49"). Like the vast majority of his fellow forty-niners, William Swain "found no gold in California," recounts Peter Coyote over a photograph of Swain as an old man in his seventies sitting outside his home with his wife and four grown children, "but he became the largest peach grower in Niagara county, New York." Coyote adds in the final moments of the episode that absence had taught him "to appreciate the comforts and blessings of home . . . but in the evenings on his farm when the work was done, he never tired of telling his wife and children and grandchildren about the great adventures he had had crossing the country when it and he had both been young."

Despite the bittersweet ending, the biographical account of William Swain and the larger epic tale of the gold rush which it illustrated are intercut with two other sections that effectively underscore the darker aspects of the western experience. The coming of modernity and urban expansion is the subject of "Emporium of the Pacific," which is primarily about San Francisco. J. S. Holliday describes at the outset of the sequence that "California was the Golgotha of sin" for the rest of the country. Gambling halls, prostitution, alcoholism, and even drug addiction followed rapidly in the wake of gold fever as tens of thousands of people rushed into the region. San Francisco grew from a town of 2,000 in 1849 to a city of 35,000 residents in 1850. The downside of such uncontrolled and rapid growth was the exploitation and misery experienced by many marginalized groups, such as the original Hispanic colonists who were largely displaced, as well as the Chinese, Native Americans, blacks, and women forced into prostitution. All of these people shared little in the half-billion dollars of gold found in California between 1849 and 1860. As early as 1852, "big [mining] machinery required big money," summarizes Peter Coyote at the end of this section, "California's gold fields were soon controlled by investors with headquarters in San Francisco, and worked by miners who worked for a weekly paycheck."

The most electrifying sequence by far in Episode Three is "The Right of Conquest," which strikes like a bombshell approximately 48 minutes into the gold rush narrative. This 7-minute 20-second scene illustrates concisely in narrative form why "the gold rush had proved a disaster for the Indians of the Plains," reverberating throughout the remaining half hour of the episode with its dramatic power and its full implications for future Native and Anglo-American relations. Ives and his creative team begin this section by raising, critiquing, and finally replacing the current and widely held misconception that Native Americans lived in relative peace and harmony for centuries before the westward expansion of whites. Anglo-American emigrants certainly brought cholera and drove away the buffalo, but many of the Plains tribes are shown in the documentary to have been in bitter rivalry with each other for generations over the acquisition of land and game. "It was not all peace before the white man came," contends Ives, "there was tremendous conflict. And if you were a Kiowa or Cheyenne living in and around the Black Hills in the eighteenth century, you were a lot more concerned about the Lakota coming west than about any Europeans."[124] Lakota anthropologist Jo Allyn Archambault reinforces this point of view with her onscreen commentary: "We moved into these plains 300 years ago. We were terrific warriors . . . [and] we swept the enemy aside and we took the land for ourselves. We took the Black Hills. We chased the other people out . . . we own those hills partly by right of conquest, and Americans understand the right of conquest."

As he does throughout *The West*, Ives puts his own personal stamp on this scene by improvising on Burns's usual documentary stylistics. With the help of supervising editor Paul Barnes, he constructs a complex combination of dissolves in which period photographs of several Indians from different tribes are superimposed over a fast-moving traveling shot taken from an airplane that visually and viscerally illuminates Richard White's troubled description of 1851 as "a period of incredible chaos on the Plains." He also incorporates more contemporary background music and sounds than are characteristic of Burns's core style, merging various instruments and traditions, especially from an assortment of Native American singers and musicians. This score is particularly effective in the scene's climactic vignette in which a needless but telling confrontation occurs between 30 soldiers and a Brulé Lakota tribe headed by Conquering Bear, the chief whom army officials had chosen to sign the 1851 Fort Laramie peace treaty between the United States and the entire Lakota nation (with other signatories for the Crows and Pawnees).

In August 1854, the Brulé had returned to Fort Laramie (in present-day Wyoming) to acquire supplies when a young cow from a Mormon wagon train strayed into the Indian camp, where it was killed with a bow and arrow. The owner of the calf complained to the fort commander who sent in troops with two howitzers to arrest the warrior who had killed the animal. In a voice slightly distorted by echo, Peter Coyote relays what happened next over a photograph of Conquering Bear. The chief, he says, "apologized and promised to pay the owner more than the animal was worth . . . suddenly the officer in charge ordered his men to fire. Conquering Bear was the first to die." The outraged Lakota then turned on the soldiers, killing all but one who escaped back to the fort before dying too. White concludes over a foreboding photograph of four Lakota warriors standing and staring straight into the camera while a native singer chants defiantly on the back of the sound track: "They killed the agent of their own relationship. They killed Conquering Bear. And what becomes a minor dispute over a cow, now becomes a sign to the Lakota that how can you trust these people."

Not every episode, sequence, and scene in *The West* is as tightly structured and as dramatically compelling as these examples. Still, the 12-hour 30-minute narrative is a strong first step toward integrating the older foundation ritual with the newer revisionist model of western history and fiction. "I'm grateful for this series and proud to be part of it," volunteered N. Scott Momaday afterward, "I think its approach is wonderful and I'm completely satisfied with what everyone has done. I, who have spent much of my life in the West and studying the West, learned a great deal from this project. I think the public will have to come away from this series with a more accurate impression of the West and its history, a much better understanding of who we are and what happened."[125] Ramón Gutiérrez wrote that this "program is destined to teach more history to Americans than all the stacks of books that gather dust in our local libraries."[126] As a final sign of institutional approval, the Organization of American Historians awarded Stephen Ives and *The West* its 1997 Erik Barnouw Prize for Outstanding Programming dealing with the study and promotion of American history.

Ken Burns's college ambition to become a successful historical documentarian came true not only for him but also for many of his friends and associates.

He clearly was a mentor to Stephen Ives, who is seven years his junior, as Ives readily credits Burns with "inspiring my decision to go into historical filmmaking, because I admired his work and I've learned a great deal from watching and being part of his process."[127] The same can be said of his younger brother, Ric, who had pursued a doctorate in English and comparative literature at Columbia University before being hired by Ken as an assistant scriptwriter and editing consultant on *The Statue of Liberty* and *Huey Long*. Ric eventually changed career tracks during the production phase of *The Civil War,* emerging from his brother's shadow in the early 1990s to establish himself as one of the leading nonfiction historical filmmakers in the country. Many other members of the extended Florentine family who worked on *Baseball* and *The West* were also caught up in Ken's enthusiasms and the projects they engendered, including most prominently Geoff Ward, Paul Barnes, Mike Hill, Lynn Novick, and Dayton Duncan. Some of these individuals would clearly have been involved in film and television production or historical writing on their own, but Ken Burns's ability and compulsion to share his dream with others affected them as well as many in the American viewing public. After wrapping *Baseball,* for instance, Burns told a journalist from *Life* magazine: "I hit the road. I go out and promote. I don't mind it. I'm good at it. I'm also thinking of taking a vacation, and it scares me to death. I've never had one. I've never been away from the studio that long. Making films like these, with people like this—it's what I've always dreamed of doing."[128]

Ken Burns's double obsession with filmmaking and history has also had a profound impact on a generation of documentary producers and directors who only know him through his work. When Burns's first few films were released in the early to mid-1980s, his techniques, though stylistically derivative, appeared unusually fresh after 25 years of preeminence and institutional control by the proponents of cinema verité and direct cinema. In the subsequent two decades, Burns's approach became the style of choice for most historical documentaries, on TV in particular. It was even copied on an occasional television commercial and parodied on prime-time cable programming such as in Nickelodeon's *Nick at Nite* ersatz documentary, *Brady: An American Chronicle* (1995), which spoofed *The Civil War* and *Baseball* by using Burns's familiar look to humorously explore how *The Brady Bunch* (ABC, 1969-1974) influences and reflects American culture. Such expressions of widespread popular recognition and acceptance are now relatively routine for Burns and his work, even though the sillier and more commercial examples can still raise concerns in the scholarly community about the way that history as a whole is currently being presented on TV.

Four and one-half years in the making, *The West* premiered on successive evenings from September 15 to 19 and from September 22 to 24, 1996, averaging slightly double the usual prime-time audience for public television. This miniseries became another GM Mark of Excellence Presentation for PBS. In conjunction with the series was the first ever GM/PBS sponsored educational website, entitled "New Perspectives on *The West*," targeted specifically at college and high school students, and featuring people, places, events, themes, archives, and a tour of *The West*.[129] A synergistic advertising and marketing campaign was also enacted by Time-Warner which had purchased the rights to all the main ancillary tie-ins. The principal book written by Geoff Ward with Stephen Ives and Ken Burns entitled *The West: An Illustrated History* was purchased in 1992 by Time-Warner's subsidiary Little, Brown for $3 million after two days of competitive bidding among five major publishers.[130] Little, Brown additionally published three related juvenile titles: *The Gold Rush* by Liza Ketchum; *People of the West* by Dayton Duncan; and *The West: An Illustrated History for Children*, also by Duncan, which was selected by the *New Yorker* as one of the 16 best children's books of 1996 and won The Wrangler award from the National Cowboy Hall of Fame that year as the best book for juveniles.

Although Time-Life Video was originally slated to be one of the major underwriters of *The West*, Burns and Ives were forced to decline their participation "when they required final creative, intellectual, and artistic control" of the series to ensure its commercial viability.[131] The producers, instead, raised the $6,863,959 budget through major grants from General Motors, the National Endowment for the Humanities, the Corporation for Public Broadcasting, PBS, and the Arthur Vining Davis Foundation.[132] Another Time-Warner division, Turner Home Entertainment, distributed the videotape version as it had with *Baseball*, promoting *The West* on radio, as well as on CNN, CNN Headline News, WTBS, TNT, and in such print outlets as the *New Yorker, People, Entertainment Weekly*, and *Esquire*. *The West*, overall, emerged as yet another major programming event for PBS as well as a popular and commercial success for Burns and his colleagues.

As Ric Burns insightfully noted about the apparent sweeping revival of popular history in general: "I think we're living in a time when history has reemerged as one of the popular forms of entertainment, and that's great. It sort of slept for a couple of decades, in the '60s and '70s, and now it's really back, as it was before TV when historical novels and historical movies and historical poetry and history itself were mainstays of popular culture."[133] In the years following the debut of *The Civil War*, moreover, television had become the principal means by which most Americans learned their history, much to the consternation of many traditionalists in the academy. More than anyone else, Ken Burns was being singled out during the 1990s as the most visible representative of this much larger trend, signaling an impending shift in his relations with some factions in the professional historical community.

CHAPTER SIX

American Lives:
Thomas Jefferson (1997)
and the Television Biography
as Popular History

WHO OWNS HISTORY?

What I want to tell you is that group of people out there that for the most part
the academy has ignored is as sophisticated as you are. They may not be versed
with the language and the nomenclature and the detailed specificity that you
engage in, and that in fact is your greatest peril as well as your greatest strength,
but they are finely tuned to the subtleties and nuances of history.

—*Ken Burns at the 1998 American Historical Association Conference*[1]

Professional historians have found themselves embroiled in a continuing debate
over the current state and quality of historical instruction in the United States
since the 1980s. The National Commission on Excellence in Education
published a 1983 report entitled *A Nation at Risk: The Imperative for Educational
Reform,* which maintained that the majority of students at all grade levels are
woefully deficient in even the most basic understanding of both American and
world history.[2] These findings were confirmed in a follow-up study conducted
by the then chair of the National Endowment for the Humanities, Lynne V.
Cheney, called *American Memory: A Report on the Humanities in the Nation's*

Public Schools, which similarly chronicled the diminished place of history in the country's elementary, middle, and high school curricula and placed the responsibility with educators across the country for the "deterioration in the pedagogy by which [they] teach whatever history has managed to survive."[3]

This controversy filtered past educational, governmental, and cultural opinion leaders to become a hot-button agenda item in the popular press as well. The *New York Times Magazine,* for example, ran feature articles trumpeting the "Decline and Fall of Teaching History" and "Clio Has a Problem"; *Time* explored in a cover story what kinds of history are appropriate in "Whose America?"; while the *Atlantic Monthly* asked "Why Study History?" and provided a stern warning not to "look in the leading high school history textbooks for the answer."[4]

Professional historians were simultaneously surprised to see a dramatic rise in historical programming on television even as they were experiencing a growing disinterest in their subject from students nationwide. The unexpected success of Ken Burns's *The Civil War* was merely the tip of the iceberg as scores of newly formed cable networks in the 1980s and throughout the 1990s became closely identified with documentaries in general and historical documentaries in particular. Besides PBS, recently created cable networks A&E, the Discovery Channel, the History Channel, the Learning Channel, Lifetime, CNN, and CBS's Eye on People all increased their historical output. The reasons were simple: Nonfiction is relatively cost effective to produce when compared to fictional programming (i.e., according to the latest estimates, per-hour budgets for a dramatic TV episode approximate $1 million while documentaries average $500,000 and reality-based programs $300,000); and, more importantly, many of these shows which have some historical dimension are just as popular with audiences as sitcoms, hour-long dramas, and movie reruns in syndication.[5]

The New England Foundation for the Humanities (NEFH), in reaction, hosted a two-day conference entitled "Telling the Story: The Media, the Public, and American History," on April 23 and 24, 1993, for the stated purpose of closely examining "the phenomenal public response to Ken Burns' public television series, 'The Civil War,'" according to JoAnna Baldwin Mallory, then executive director of NEFH, and the "truly astonishing work, a fluorescence of documentary filmmaking and historical programming that has come to national attention . . . in a mere decade."[6] Clearly this wellspring of new made-for-TV histories was attracting not only large audiences but also the notice of the professional historical establishment. In a recent article in *Public Historian,* Gerald Herman, a history professor at Northeastern University with extensive media production experience, recalled how "most historians for a long time insisted on the marginality of [film and television] presentations to their concerns, to their training, to their individualized methods of work."[7] He added that "respected historians didn't bother to list their work with media-based presentations on their *curriculum vitae* for fear of having their reputations as serious scholars diminished by the association."[8]

In contrast to this attitude, a small but committed group of scholars led by John O'Connor and Martin Jackson formed the Historians Film Committee at the 1970 American Historical Association annual conference, publishing its first issue of *Film & History* the following year. Still, interest in "film and television as historian" only reached a critical mass among professional historians in the mid-1980s, largely in response to the marked rise in popular mediated productions on historical subjects that emanated from both inside and, most surprisingly, outside the academy. Because of the unprecedented response to *The Civil War* as well as the widespread attention and accolades accorded his other work, Ken Burns emerged as the signature figure for this nascent historical documentary movement. He, in turn, became a lightning rod for scholars to express a spectrum of pro and con reactions about the growing popularity of films and television programs about the past that overshadowed the former preeminence of written histories alone. The mostly complementary relations that Burns had typically enjoyed with the academy up to that point were now proving far more complicated as a result of his own heightened profile and success as well as those of the historical documentary.

In 1997, historian Joseph Ellis coined the term "pastism," to refer to the "scholarly tendency to declare the past off limits to nonscholars."[9] Robert Sklar perfectly captured this longstanding bias in the context of film and history with his metaphor "historian-cop," which alludes to the tone of policing that usually emerges whenever academic historians apply the standards they reserve for scholarly books and articles to motion pictures. Sklar, in this specific instance, called for a greater awareness of both the production and reception processes of filmmaking as a better way of appreciating how these more encompassing frameworks influence what audiences actually see and understand as history on the screen.[10]

History on television is an even more tempting and incendiary target than film as history for the proponents of pastism, especially since the impact and popularity of TV with the general public far outstrips anything that can ever be achieved in theaters. As a result, made-for-television histories are sometimes rejected out of hand for being either too biographical or quasi-biographical in approach or too stylized and unrealistic in their plot structures and imagery.

Occasionally these criticisms are well founded; historical programming certainly furnishes its share of honest failures or downright irresponsible and trashy depictions of the past. Other times, though, television as historian delivers ably on its potential as popular history, as is the case with the skillfully constructed and informative *The Civil War*.

The historic transformations from oral testimony to writing to print to film and television are generally understood today as producing concurrent shifts in the way societies privilege certain forms of expression and knowledge over others. Some historians, for example, have admonished Burns for emphasizing the empathetic and experiential aspects of history in *The Civil War* more than detailed analysis.[11] What such criticisms overlook, however, is that the visual media's codes of historical representation are far different from, though often complementary to, those of print. The present image-based histories of Burns and other producer-directors feature the simulated experience of being there by engulfing viewers in a sense of *immediacy,* in contrast to the printed word's propensity toward logic, detachment, and reasoned discourse.

Professional history typically rejects the mythmaking of popular history. This tradition, which dates back to the second half of the nineteenth century, recasts the study of history inside the larger framework of scientific inquiry with an allegiance to objectivity (albeit modified these days), a systematic and detached method of investigation, and the pursuit of new knowledge. The much older legacy of popular history, in contrast, is far more artistic and ceremonial in approach. It is usually consensus-oriented, narrative and biographical in structure, and intended to link producers and audiences in a mainly affirming relationship based on the immediate experience they are sharing together around the characters and events of their cultural past. The most prominent and influential examples of popular history in America are now surprisingly found on prime-time television, a medium from which historical programming was largely absent throughout much of the 1970s and into the 1980s. Ken Burns has come to symbolize this phenomenon in many people's minds, mostly on account of the unprecedented impact and reception garnered by *The Civil War* and *Baseball*.

Criticisms of *The Civil War* and other historical documentaries also rest on differences of interpretation, of course, although a more fundamental struggle over authority and control of historical activity in general is never too far from the surface of these present-day debates between professional historians and their popular counterparts. Valuable critiques of *The Civil War* concerning its cursory portrayal of Reconstruction, for instance, or the need for a fuller representation of the role played by African Americans in bringing about the social transformations now associated with the conflict were both raised at the 1993 NEFH Conference. These two critiques along with five other responses to the series, ranging from complimentary to ambivalent to condemning, were later published in *Ken Burns's The Civil War: Historians Respond.*[12] Occasionally in these debates, spoken remarks or written passages slip into the actual analyses that disclose as much about the deeper concerns of some academic critics as about Burns and *The Civil War.* For example, Catherine Clinton in her essay, "Noble Women as Well" calls Burns a "documentary *Wunderkind* who has rejuvenated serious interest in history—from networks, corporations, and perhaps, even the viewing public. Burns's historical influence has brought people back to reading (or at least buying!) more books, created a vogue in Civil War scholarship (especially for the new media darling, Shelby Foote)." Yet Clinton also reveals, "I regret it is so self-serving and ironic for me to trash Burns on the topic of his egregious and blatant neglect of women, as two of my last three books are shaped to deal with the topic of women and the Civil War . . . But after being drafted, it is *not* a tough job, and somebody's got to do it."[13]

A recent article in the Chronicle of Higher Education was pointedly titled "Taking Aim at the 'Ken Burns' View of the Civil War."[14] Historian Edward Ayers provided a revised interpretation that rejected the "established narrative" imbuing the war with "an overarching moral purpose it lacked at the time." Ayers continued, "since the dawn of the civil-rights movement . . . historians have oversimplified the war," singling out James M. McPherson, author of the highly regarded *Battle Cry of Freedom,* as a prime example.[15] Most tellingly, a generation of accepted historical thinking is not characterized here as the "McPherson View of the Civil War," but as the "Ken Burns View," reflecting both the *de rigueur* dismissal of an outlook that has been turned into popular televised history, as well as the unstated though implied recognition that *The Civil War* was built upon the foundation of academic scholarship, however conventional that historical record may be from this new revisionist point of view.

All of Burns's historical documentaries, in fact, are modeled on existing research and designed to bridge public interest in the subjects he chooses with the findings of the scholarly community. *Thomas Jefferson,* for example, incorporated roughly a quarter century of professional historical thinking on its subject, while it also attracting a reported 17 million viewers when it debuted on public television on February 18 and 19, 1997.[16] On the more than 1,000-page interactive website created to accompany the miniseries, Burns discussed his role as a popularizer of academic ideas: "We are in the business of helping to disseminate ideas—challenging ideas, contradictory ideas, tragic ideas, powerful ideas. And Jefferson is of course a master at all of those things. So we're looking for those scholars who can help us set into vibration the facts of his life with the ideas of his life."[17] *Thomas Jefferson,* the premiere episode of the *American Lives* series, is richly indicative of Ken Burns's general approach to the television biography as popular history. As he further describes his process and perspective involving this specific subject: "You set out with a desire to learn about Thomas Jefferson and in the course of things you enrich yourself by that process of discovery . . . I go at it looking for Thomas Jefferson and the Thomas Jefferson that I found is not THE Thomas Jefferson, but my Thomas Jefferson."[18]

MEDIATING PAST AND PRESENT

Adams wrote back to Jefferson and said, "We ought not to die before we have explained ourselves to each other." We Americans ought not to die either before we have explained Thomas Jefferson to each other, before we have come to terms with the protean genius and completely human man who wrote the words that form the heart and soul of our great nation.

—*Ken Burns, 1997*[19]

Ken Burns is an "ideal filmmaker for this period of transition between generations, bridging the sensibilities of the people who came of age during World War II along with his own frame of reference as a baby boomer."[20] He explores America's heritage in the subjects he selects, responding to those aspects of the past that he and his colleagues find most relevant and compelling, while leaving behind that which is nonessential to present day concerns. "All of the contradictions in Thomas Jefferson's life and times," he says, "are played out again in our late twentieth-century national life."[21] According to Burns, Jefferson "helps define the issues which will animate our national discourse right up to the present."[22] This artistic reintegration of the past into the present is one of the major functions of popular history. It is a process of reevaluating the country's historical legacy and reconfirming it from a new generational perspective. As Burns reveals:

I now think that these subjects choose me. I have been saying for a very long time that I was interested in a very simple question, which is "Who are we?" That is to say, what does an investigation of the past tell us, Americans, about who we are . . . I now begin to think that I am very much tied up in the asking of that question—that these films are also a way of saying, "Who am I?" And that gets a little bit more confusing and harder to nail down. Suffice to say, I think that these projects are chosen because they are compelling dramatic stories. They are chosen because they have as a central feature an element of biography . . . They are driven by the notion that people can change events— that people do change events for the better and for worse.[23]

In the case of *Thomas Jefferson*, Burns engages a subject who is much more a figure of words and ideas than of physical action. The filmmaker dramatizes the themes associated with Jefferson by the narrative choices he makes as well as by his usual strategy of employing expert commentators to personalize the concepts being presented. In assembling his plot structure, Burns uses the chronological events of Jefferson's life as the fundamental storyline on which to anchor the narrative. In Part I especially, Jefferson's family history is used to humanize Thomas Jefferson, the icon; and family history similarly becomes another bridge to national history and the discussion of Jefferson and race, Jefferson and the role of government, and Jefferson and the meaning of freedom, the three most important issues in the series.

Burns and chief writer Geoff Ward construct the plot, for example, around the three most celebrated concepts in the Declaration of Independence, "Life," "Liberty," and the "Pursuit of Happiness." Part I, which is 87 minutes long, introduces Jefferson as a young son and bookish student in "Life," and as a husband and rising political star in "Liberty: Our Sacred Honor." This portion of the miniseries contains a far greater number of viewer involvement strategies than does Part II, such as the many opportunities to identify with Jefferson falling in love, honeymooning at Monticello, and becoming a father. These intimate moments heighten the accumulated effect of the historically significant scenes, for instance the first-person camera point of view when the audience is literally placed inside Jefferson's room with him in Philadelphia during the writing of the Declaration of Independence.

The most arresting personal vignette occurs approximately one hour into the program when Sam Waterston and Blythe Danner read passages from Laurence Sterne's *Tristram Shandy* in voice-over to simulate the loving interaction between Jefferson and wife, Martha, nicknamed "Patty," who is dying after a difficult childbirth. As this exchange intensifies, the rhythms of the actors and the pacing of the interior live shots inside the bedroom at Monticello almost function like a traditional dramatic scene. Later toward the end of Part I, there is another five-minute scene involving the love affair between Jefferson and Maria Cosway in Paris, where Burns noticeably begins to shift dramatic gears. Waterston performs an affecting reading of Jefferson's famous "Head and Heart" letter, but, significantly, his first-person monologue stands in stark contrast to the two-person interplay of the earlier *Tristram Shandy* scene, since Maria Cosway has now returned to her home in England. The camera holds for a long take on a lovely painting of Maria, but Jefferson is basically left to himself, trying to sort out his thoughts and feelings alone. The effect is to intimate that Thomas Jefferson, the character, is emotionally withdrawing, just as onscreen commentator Clay Jenkinson concludes that this affair "produced a crisis for Jefferson . . . [who] reasserted the head . . . fe[eling] that human relations were too painful and that it was simply better to live in a world of abstraction and ideas and architecture," thus prefiguring the more conceptual agenda of Part II.

The second half of *Thomas Jefferson,* lasting 89 minutes, is far less personal and more contemplative than Part I. There are a few exceptions, such as the emotional exchange of letters between Jefferson and John Adams, although even this sequence, "The Pursuit of Happiness," is much more discursive in structure than the earlier dramatic portions of the program. Ken Burns is far more interested in words than most filmmakers, and he purposely slows down the pace of Part II to more fully explore the principal themes introduced earlier in the series. Burns's handling of the Sally Hemings controversy is a case in point. This scene takes place during "Liberty: The Age of Experiment" in Part II, a sequence that examines Jefferson's terms as secretary of state, vice president, and president. Hemings is introduced in a Federalist broadside in 1802, claiming that this young slave of Jefferson is also his long-time mistress and the mother of his mulatto son, Tom. Burns investigates this charge and the related issues of race, slavery, and freedom by constructing an editing cluster, which involves cutting together images of his subjects with a montage of commentators who typically present both corroborating and conflicting opinions, creating a collage of multiple viewpoints.

In this 5-minute 45-second scene, shots of Jefferson (Rembrandt Peale's 1800 painting) and Monticello (both interior and exterior photographs) are intercut with five differing reactions to the controversy: Clay Jenkinson ("We don't know. The evidence is slender."); Natalie Bober ("a moral impossibility . . ."); Robert Cooley, Hemings descendant ("I have the benefit of 200 years of consistent, solid, oral history . . . Sally, was without a doubt, his mistress, lover, and substitute wife for 38 years."); Joseph Ellis ("If it were a legal case . . . the evidence would now be such that Jefferson would be found not guilty."); and John Hope Franklin ("It doesn't really matter whether he slept with her or not. He could have. After all he owned her. She was subject to his exploitation in every conceivable way."). Burns's own position is readily apparent by the editing choices he makes which lead inevitably and inexorably to John

Hope Franklin, who forcefully articulates the broader context and implications of the controversy during the final minute and 15 seconds of the scene: Thomas Jefferson owned many slaves, providing him with a privileged existence at their expense. He is, therefore, guilty of profiting by and supporting an institution that allowed other white masters all over the south to sleep with their slaves, whether or not he himself was ever intimately involved with Sally Hemings.

Ken Burns, moreover, follows the example found in much of the historical literature by identifying Jefferson with an assortment of current issues of national interest, especially in Part II. Joseph Ellis traces this tendency back to nineteenth-century historian James Parton, whom he quotes in *American Sphinx: The Character of Thomas Jefferson:* "'If Jefferson was wrong, America is wrong. If America is right, Jefferson was right.'"[24] Ellis emphasizes in his book how many eminent professional historians from the past, including Jefferson biographers Dumas Malone and Merrill Peterson, as well as scores of current citizens from all walks of life still evoke Jefferson as a way of discussing American culture and society.[25] Burns's approach in *Thomas Jefferson* is to similarly envision Jefferson as "a kind of Rosetta Stone of the American experience."[26] He explains:

> When we talk about the separation of church and state, prayer in the classroom, school funding for parochial education, Thomas Jefferson is there looking over our shoulders. When we debate states rights versus big government and think about the tension between home-grown militias on the one hand and a monolithic federal government on the other . . . When we think about the intractable problems in our country born of race . . . Thomas Jefferson and his agonizing internal contradictions are looking over our shoulder making us who we are for better or worse. [27]

As a way of better integrating these issues into the plot, Burns uses Clay Jenkinson almost as a second surrogate narrator to complement Ossie Davis throughout Part II. The filmmaker utilizes him 20 times, or more than half of the 38 total commentaries in this entire final half of *Thomas Jefferson*. Jenkinson, a National Humanities Medal–winning Jefferson impersonator and professor at the University of Nevada at Reno, provides plenty of anecdotes to animate Part II's emphasis on "abstraction and ideas and architecture." He, too, is a skilled popular historian, contributing a rich human interest dimension to the miniseries, even though he never assumes the character of Jefferson in the film. Jenkinson, instead, offers many background stories which link Jefferson's private life to his cultural interests and his political ideas. The camera even enters into Thomas Jefferson's inner sanctum in Part II, dissolving through a montage of ethereal black-and-white interior shots of hallways, rooms, and furnishings, as Jenkinson suggests onscreen that "Monticello is most of all a metaphor for Jefferson's soul." In effect, Ken Burns directs his undivided attention onto a single individual of consequence from the past throughout *Thomas Jefferson*, striving to stimulate for himself and his audience the kind of intense connection with a historical character that is usually only achieved in fiction filmmaking.

MEDIATING OBJECTIVE AND SUBJECTIVE STYLISTICS

A documentary has as much artistic possibility as a fiction film . . . history is just the medium, like a painter choosing oil as opposed to watercolors, that's what I work in, but, first and foremost, I am a filmmaker trying to learn my craft.

—*Ken Burns, 1997* [28]

The documentary narrative is a particular mode of knowledge and means of relaying history, and Ken Burns uses the inherent characteristics of photography, film, and television to create his popular histories. Burns, like many other producer-directors of his generation, is often preoccupied with traversing the stylistic border between fact and fiction. The pre-photography nature of *Thomas Jefferson* actually induced him to experiment much more with the documentary form than he usually does, given his largely traditional approach to media form and aesthetics. Burns, for example, commissioned architectural photographer Robert C. Lautman to take hundreds of platinum palladium prints with a nineteenth-century view camera inside and outside Monticello and throughout the accompanying slave quarters, so he could approximate the look of old archival images which, of course, do not exist from Jefferson's lifetime.[29] The filmmaker's intention was to rephotograph Lautman's stills, thus realizing one of his main strategies in this documentary of portraying Monticello as a visual analogy for Jefferson himself while also continuing one of his trademark techniques.

Burns typically reshoots photographs as if they were moving pictures, panning and zooming within the frame, shifting between long shots, medium shots, and close-ups, turning these single images into scenes rather than just shots. In the final 20 minutes of *Thomas Jefferson*, for example, there is a brief, bittersweet 45-second scene composed entirely of one of Robert Lautman's antique looking stills, rephotographed from three vantage points. As Sam Waterston reads a portion of one of Jefferson's last letters to John Adams, the camera shows his writing table situated near his alcove bed; there follows a cut-in of the tabletop as Waterston's voice-over recollects "when youth and health made happiness out of everything"; and, finally, the vignette climaxes with a close-up of the seat where Jefferson once sat as his spoken words intimately share with Adams the calm realization that they are both close to "the friendly hand of death."

Burns additionally recruited Peter Hutton to produce time-lapsed black-and-white film footage of the interior of Monticello to intercut with his rephotographed images.[30] Hutton's camerawork is featured prominently in the many montages of Monticello throughout Part II, as well as in the aforementioned *Tristram Shandy* scene in Part I. Ken Burns characteristically intercuts the highly active rephotographed footage with Hutton's live shots which are framed more like still photographs, thus simulating the mood and pre-film vocabulary of the late eighteenth and early nineteenth centuries of the subject under review. Burns describes this expansion of his technical and grammatical repertoire as "eighteenth-century virtual reality."[31] In point of fact, his stylistic approach to documenting reality is far less subjective than many of his contemporaries. In a recent round table discussion on the state of the documentary, thirteen of Burns's peers generally agreed that the line between nonfiction and fiction "is an illusory distinction."[32] Ken Burns, too, mediates the stylistic distinctions between fact and fiction, although he relies mostly on techniques introduced decades ago, such as rephotographing and time-lapse cinematography, which are both a half-century old.

Even Burns's inclusion of visual reenactments for the first time in his career in *Thomas Jefferson* was similarly understated in its application. These three shots, lasting less than 30 seconds each, take the form of either a horse-drawn coach or a man alone on horseback silhouetted against the pink twilight. They occur during the introduction, following the death of Patty Jefferson, and on Jefferson's return to Monticello after the completion of his presidency. They are all intended to lyrically suggest the presence of the protagonist, therefore, once again, expanding the available imagery. Despite the controversial nature of reenactments these days, Burns has aurally employed this strategy since the beginning of his career by his "chorus of voices" technique—his use of actors and actresses to deliver dramatic readings from diaries, letters, personal papers, and other printed recollections of various kinds. Burns regularly integrates live and historical source material, putting each on an equal footing in the present tense, thus rendering these subjects from the past more accessible and immediate to modern audiences. The aesthetic effect he is striving for is to "bring the past back alive onscreen."[33]

All told, Ken Burns's historical documentary style is poetic realism, capitalizing on the inherent ability of photography, film, and television to suggest analogies (e.g., the many sides of Jefferson are intimated by the architecture, furnishings, and grounds at Monticello; the self-divisions in Jefferson's personality are reflective of the nation as a whole) more than to assert precise meanings (which, of course, is a basic strength of written discourse). As a result, Burns's made-for-TV history of Jefferson portrays the contradictions in his character, debates them, but never provides any final resolutions. The ambiguities that reside in *Thomas Jefferson* and all of his other television histories, moreover, afford audiences of tens of millions some interpretive space in which to explore differing ideas and opinions, and most essentially, to engage with figures like Jefferson and his times in the present, which is the penultimate goal of popular history. In Burns's own words, "we're not here to debate as much as we are to cohere."[34] His documentary style, in turn, expresses his liberal pluralist leanings, offering a view of the United States that is basically fixed on agreement and unity, even as it struggles with its heritage of race and slavery and the place of Thomas Jefferson in contemporary life.

MEDIATING IDEOLOGICAL DIFFERENCES

Unfortunately and tragically I would say that, in a sense, Thomas Jefferson personifies the United States and its history. He was a man who claimed to be a man of the Enlightenment. He was a scientist, a humanist. He knew what he was saying when he said that all men are created equal. And it simply can't be reconciled with the institution of slavery.

—*John Hope Franklin in Thomas Jefferson*[35]

Most of Ken Burns's subjects are majoritarian rather than marginalized— for example the Statue of Liberty, the Congress, the Civil War, baseball, and, of course, Thomas Jefferson—although he does incorporate multicultural issues and outlooks into the broader panorama of his nationalist narratives. As discussed earlier, Burns has always contended that by making these documentaries he is "asking one deceptively simple question: Who are we? That is to say, who are we Americans as a people?"[36] This preoccupation with the elemental question—"Who are we as Americans?"—could not be more relevant in an era when multiculturalism has become the source of sweeping and fundamental reappraisals of almost every aspect of national life. *Thomas Jefferson* is designed as such a reexamination. Jefferson's image is clearly in transition today, and his racial legacy is the major reason why he now occupies such a problematic place in American history and culture.

Burns's most effective tool in reexamining Jefferson's meaning in the present is, once again, his editing clusters, his linking together Jefferson-related imagery with a montage of assorted commentaries. In the coda, for example, paintings of Jefferson by Rembrandt Peale (1800), Charles Wilson Peale (1791), and Gilbert Stuart (1805) are interspersed with seven separate opinions of Jefferson's accomplishments, his shortcomings, and his current significance—Joseph Ellis: "There is a simple but extraordinarily resonant message that Jefferson somehow symbolizes, namely the future is going to be better than the past"; Gary Wills: "I think the thing to remember from Jefferson is the power of the word—that ideas matter"; a shot of the Declaration of Independence pans across the phrase: "life, liberty, and the pursuit of happiness"; Clay Jenkinson: "It is Jefferson who is indispensable because he is mysterious, idealistic, pragmatic, misunderstood, complicated, paradoxical, hypocritical. He is the stuff of America and that is who we are and that is why Jefferson has to be the center of our national discourse"; a shot of a slave; John Hope Franklin: "The legacy of Jefferson is both a gift and a curse . . . he cursed us with a practice of inequality and slavery and a denial of justice that scarcely can be erased by anything we can think of."

Burns, then, allows the audience to rest for a moment and absorb what's been said as he intercuts an old photograph of Monticello, the Capitol in Washington, D.C., at midcentury, another image of several slaves, and a live shot of the Jefferson Memorial before Andrew Burstein prefigures the Civil War in the next statement, "I don't think he was convinced that America would be able to advance without fits and seizures and numerous torments. He didn't know how to hold the union together, but in the end I'm sure he felt he had done his best—that he had lived up to his dreams." The coda continues with Gore Vidal: "With all his faults and contradictions . . . if there is such a thing as an American spirit, then he is it." Finally, Clay Jenkinson returns offscreen over various portraits of Jefferson:

> Jefferson essentially tells us that we cannot be complacent until two conditions are met. Every human being born on this continent has a right to equal, indeed, identical treatment in the machine of the law, irrespective of race, gender, creed, or class of origin. And, secondly, everyone born on this continent has a right to roughly equal opportunity at modest prosperity, and until these conditions are met, we cannot rest. When those conditions are met, we may say as Jefferson said he would, *nunc dimittis,* you may dismiss me, my work is done.

In summary, therefore, individual speakers differ on the exact meaning of Jefferson's legacy throughout this editing cluster, but disagreement ultimately takes places within the broader framework of agreement on underlying principle. The scene ends with a dramatic time-lapse shot of the sun setting with Sam Waterston speaking the words of Thomas Jefferson about the enduring nature of representative government and "this country['s aim] to preserve and restore life and liberty."

Ken Burns, overall, articulates a version of the country's past that conveys his own perspective as a popular historian, intermingling many widespread assumptions about the character of America and its liberal pluralist aspirations. Like other documentarians of his generation, he, too, addresses matters of diversity, but unlike many of his contemporaries, he presents an image of the United States pulling together despite its chronic differences rather than a society coming apart at the seams. In his own words, "I know I've said it before but I see myself as an emotional archaeologist, trying to excavate what there is in our history that speaks to the *unum* and not the *pluribus*."[37] Exploring the past is also his way of reassembling a future from a fragmented present. Clay Jenkinson's final commentary, in particular, reminds viewers that much work still needs to be done before Americans can more fully enact the essential ideals that Jefferson professed.

FINDING A PLACE FOR POPULAR HISTORY
ALONGSIDE PROFESSIONAL HISTORY

Thomas Jefferson is the most intensely personal film I've made.
—*Ken Burns, 1998*[38]

During the six week promotional tour preceding the debut telecast of *Thomas Jefferson,* Ken Burns gave literally hundreds of interviews, delivered dozens of variations of his prepared speech, "Searching for Thomas Jefferson," and screened portions of a selected clip from the series whenever the opportunity arose. The brief segment Burns selected to show in this context was the writing of the Declaration of Independence scene from Part I. This entire 8-minute 30-second set piece is skillfully executed, beginning with the activities at the Continental Congress where Jefferson is assigned the job and culminating inside the small room at Philadelphia where he actually completed the submitted draft of the famous document. One minute into the scene, Clay Jenkinson tellingly portrays the personality of Thomas Jefferson as "bland and careful and aphoristic and high flown, his rhetoric always soared toward aspiration and human dignity." This characterization also suggests a certain similarity to Ken Burns's poetic stylistics and his empathetic (and sometimes romantic) approach to his material, indicating in part why this producer-director identified so closely with this particular subject in such an intensely personal way.

All of Burns's work demonstrates certain ideological, narrative, and biographical imperatives which support one another and together form an image of America that is primarily democratic in outlook, dramatic in structure, and intimate in portrayal (i.e., "his goal . . . was to explore the 'inner Jefferson.'").[39] The nature of the historical biography is yet another reason why Burns and a viewership of millions responded so personally to the Thomas Jefferson who emerged in this miniseries. Ken Burns formed a strong attachment to his subject over a seven-year period while overseeing this project from initial concept to finished three-hour documentary as its executive producer-director.[40] Audience members, too, make their own kind of individual commitment during two 90-minute viewing sessions which simulate powerful feelings of intimacy for them as they watch and relate to the featured character's life story on TV. This interlocking ritual of producing, telecasting, and watching *Thomas Jefferson* becomes a shared ceremonial experience for both the filmmakers as well as the vast numbers of Americans who tune in to see this newly adapted screen version of the historical Jefferson.

Monticello is again an apt analogy for this intimate biography. Jack McLaughlin observes: "There is no denying that Monticello has become a Jefferson museum-shrine, but most examples of restored domestic architecture are museums that lift the past out of context and place it on display. An inhabited house, unlike a museum, bears the imprint of its owners: the ashes in the hearth, the fingerprints on the walls, the scuffs, knocks, scrapes, and rubbings of human contact—the detritus of everyday life. All of these vanish beneath the cosmetic touch of the restorer's art."[41] Ken Burns's art involves similar slights of hand. The stylistic features of photography, film, and television strongly influence and embellish the kinds of historical representations that he and his colleagues create. Camera reality in *Thomas Jefferson* is comparably revivifying and pristine, much like the condition of the living preserve at Monticello—a highly decorative, visual tribute to Jefferson, only this time shot on film. Burns and his crew employ their considerable formal talents as artists and popular historians to raise a semblance of Jefferson from the past and insert him into the present tense of television, offering a prime-time special event for literally millions to see, hear, and, most importantly, identify with in the privacy and comfort of their own homes.

Made-for-TV histories are, thus, never conceived according to the standards of professional history. They are not intended chiefly to debate issues, challenge the conventional wisdom, and create new knowledge and perspectives. *Thomas Jefferson*, specifically, is designed for the far less contentious environment of public television, supported largely by the continuing patronage of well-established governmental and corporate sponsors.[42] As a result, Burns's popular history is artfully serious and respectful, warmly and sumptuously photographed, and occasionally rhapsodic in tone. Part II of *Thomas Jefferson*, for instance, begins with George Will's reverential appraisal: "Jefferson was, I think, the man of this millennium. The story of this millennium is the gradual expansion of freedom and an expanding exclusion of variously excluded groups. He exemplified in his life what a free person ought to look like. That is someone restless and questing his whole life under the rigorous discipline of freedom."

As a corrective to such grandiloquent remarks, Burns also injects more critical assessments, fashioning a more usable and realistic Jefferson for a contemporary America struggling anew with the challenges of race and diversity. Historian Paul Finkleman's counterpoint provides an example:

> One of the defenses of Jefferson is "well, he was just a Virginia planter and we can't expect anything else from him. He was just like his neighbors." And I think the point to be made is that he's not just like his neighbors. We don't build monuments to people who are just like their neighbors. We don't put them on the nickel. We don't make them icons. Jefferson was a very special man and we expect more from him. So we compare him to the best of his generation, not merely average. We compare him to Washington who freed his slaves, to his cousin John Randolph of Roanoke who freed his slaves, to his neighbor Edward Coles, to the thousands of individual small Virginians who freed their slaves. The free black population of Virginia grows from 2,000 to 30,000 in a space of about 30 years. A lot of Virginians were freeing their slaves. Where's the master of Monticello? Why isn't he there?

The Jefferson of *Thomas Jefferson* is, therefore, a more complicated and conflicted figure than is evident in the nearly three-dozen previous film and television depictions—such as *1776* (1972), *The Adams Chronicles* (1976), *The Rebels* (1979), *Jefferson in Paris* (1995), and *LIBERTY! The American Revolution* (1997), just to name the most prominent post-1970 examples—although Burns's documentary never approximates the comprehensiveness and precision of the existing academic literature. *Thomas Jefferson* incorporates commentaries by professional historians and subsequently aids in the popularization of their scholarly work (e.g., Joseph Ellis's *American Sphinx* became a best seller after being strategically released by Knopf the same week as the television series).[43] Like other expressions of popular history, however, Burns's documentary on Jefferson is only a partial reflection of the published record. *Thomas Jefferson* and its producer-director never pretend to present all there is to know on the subject; that is neither the strength nor the purpose of photography, film, and television as history. This historical documentary, instead, is far more significant because of its ceremonial ability to connect unprecedentedly large television audiences in the present with a shared sense of their common past.

Above all else, then, Burns's popular history is an intermediary site bridging the findings of professional historians with the interests of the general public. *Thomas Jefferson* functions, first and foremost, as the focal point of a large-scale cultural ritual based on fusing the stories of the past with the concerns of the present for a vast contemporary viewership. The current controversies over Jefferson reflect the internal divisions that now exist over the very definition of America's national identity. Ceremonial historical narratives, such as Ken Burns's *Thomas Jefferson,* are artistic attempts to reconstitute the cracks and fissures of that identity at a new point of agreement and consensus. Burns's work as a whole is an artistic reimagining of the national sense of self from a new generational perspective, that is, "Who are we Americans as a people."

Scholars, too, can seize this opportunity to reach beyond the academy and engage the outside community more with their own increasingly original and detailed accounts of past events, figures, and issues. The widespread popularity of historical documentaries on TV is indeed a reminder that history is for everyone. Some histories can mediate differences and reawaken what people take for granted in their collective past. Other more revisionist and unconventional approaches to history can also challenge established values by provoking, interrogating, unsettling. *Thomas Jefferson* falls into the former category; and no one has drawn more Americans to history of every kind through the power and reach of prime-time television than Ken Burns.

PRODUCING HISTORY
FOR MILLIONS IN A POPULAR TRADITION

I've been working in two parallel tracks. One has been a trilogy of three major series— *The Civil War, Baseball,* and *Jazz . . .* And in a parallel track, I've been

working on a series of biographical portraits, necessarily smaller films, not just
because these other series take up more time, but because I wanted to work in
a smaller form.

—*Ken Burns, 1999*[44]

Biography is the central organizing principle around which Ken Burns
builds all of his historical documentaries. He believes that "[t]he way we come
to terms with our common past is through a doorway that is the lives of other
people."[45] Even *Brooklyn Bridge,* in this way, starts out as a dramatic rendering
of John and Washington Roebling's combined struggle to design and oversee
the construction of this inspired American landmark. Mother Ann Lee is
similarly the seminal presence behind the story of *The Shakers. The Statue of
Liberty* begins by chronicling the creative obsession of Frédéric Auguste Bartholdi. *The Congress* is successively comprised of historic accounts and anecdotes
concerning that institution's "Builders," "Debaters," "Bosses," "Progressives,"
and "Managers." *The Civil War*'s structural center of gravity are those nearly 50
famous as well as anonymous participants whom Burns enlists to animate his
11-hour version of the conflict. *Baseball*'s over 18-hour narrative likewise boasts
a cast of approximately 75 key figures. *Huey Long* and *Thomas Hart Benton* were
clearly his first direct attempts at producing smaller biographical films. It was
then, in the late 1980s, too, that he initially conceived of "*American Lives* [as] a
periodic series of historical documentaries on the lives of extraordinary Americans, people in many cases so well known that they have become cliched to the
point of neglect."[46]

Burns actually decided to produce and direct a biography of Thomas
Jefferson in 1983 even though he was fully preoccupied with other projects at
the time. He admits now he is "glad [he] waited" until he was older and more
mature to explore such a complex subject as Jefferson through a film he describes
"as a prequel to *The Civil War,* an autopsy performed on the country in its
infancy to understand more the pathology of a country that would come to near
national suicide four score and five years later."[47] His curiosity about Jefferson
also cultivated in Burns a growing interest in Lewis and Clark, whom he first
considered pursuing as a dual biography after reading a 1988 book entitled *Out
West: American Journey Along the Lewis and Clark Trail,* by Dayton Duncan, a
friend who was then consulting on *The Civil War.*[48] Another personal acquaintance and professional colleague led to the genesis of *Not For Ourselves Alone:*

> I remember ten years ago when I was editing *The Civil War* series in the summer
> of 1989—my editor Paul Barnes and I would have lunch together with most

of the editors every day, and he regaled us for several weeks at the progress he was making on a book about the life of Elizabeth Cady Stanton written by Elisabeth Griffith. Every time he told us stuff, I felt like I was being knocked over by what I did not know, what I hadn't heard of. Basically coming out of that series of lunches I said to Paul, "We have to make this film."[49]

While editing *The Civil War,* Burns also thought about producing a more focused and in-depth exploration of Abraham Lincoln. His head scriptwriter Geoff Ward was concurrently absorbed in writing and researching two successive biographies on the young Franklin Roosevelt.[50] "The [d]evelopment of FDR" looked to Burns to be another "fascinating . . . study" to film.[51] Ward, Duncan, and Burns, moreover, "talked about doing a documentary biography of Mark Twain, whom [they] consider[ed] one of America's greatest yet most misunderstood authors."[52] "I could do ten lifetimes of these stories if I started doing them all," Burns enthusiastically affirms about the long-term viability of his biographical series, "it would take me a thousand years just to cover the last 150 in American history."[53]

The success of *The Civil War* paved the way for his realization of *American Lives.* Burns began his relationship with General Motors in the summer of 1987. "I was thrilled," he recalls, "the support of General Motors meant not only that *The Civil War* would get done, but that it would be seen and studied, thanks to GM's mission to provide both a strong public relations program and a comprehensive educational outreach component."[54] After *The Civil War* became the highest rated miniseries in public television history, General Motors committed to being the sole corporate underwriter for Ken Burns's work under its Mark of Excellence Presentations program for the remainder of the decade, leading to its major participation in the financing and widespread promotion and distribution of *Baseball, The West,* and *Jazz.*

In the smaller biographical format, Burns also acquired follow-up monies to complete *Empire of the Air: The Men Who Made Radio,* which was well into shooting and assembly by late 1990.[55] He then proposed *American Lives* as a five-episode package budgeted between $1 and $2 million per installment, mostly as a way of jump-starting the always challenging fundraising process. When General Motors agreed to support this series, Burns essentially secured 35 percent of the cost up front rather than trying to procure all the monies separately for each individual biography.[56] In reflecting on his continuing partnership with the automaker, the producer-director explains that "[i]t has been a wonderful, sympathetic, and symbiotic relationship—a relationship that gives me the freedom of range to risk, to be bold, to try new avenues."[57] "In the beginning," he told one interviewer, "I was very suspicious . . . I was just waiting for the other shoe to drop. It's now been 12 years and it's never dropped."[58]

Ken Burns's distinctive approach to the television biography is most apparent in the areas of production-distribution, common strategies in plotting and characterization, and the contemporary relevance of his historical subjects. The influence of General Motors is evident in the first category, specifically in stabilizing Burns's budgetary prospects as well as adapting the sophisticated marketing techniques already utilized so successfully with miniseries such as *Baseball* and *The West* to these smaller format biographies, thus enhancing their presence across numerous media and, in turn, their access to a far wider audience. The producer-director frequently contends that his relationship with General Motors is "absolutely noninterfering" when it comes to the actual production process: "I don't tell them how to make cars, and they have never once told me how to make a film."[59] The automaker's contribution beyond that, however, provides Burns with a financial head start to create programming with the highest possible production values when compared to the other kinds of biographies currently on TV.

Fifteen biographical programs were thriving on U.S. television in 1999 with a half dozen more already in preparation.[60] Even though most of these existing series are among the most watched shows on their respective networks, the characteristic budgets in this TV genre usually range from an estimated $350,000 for upscale biographies at the History Channel and *Bravo Profiles* to $250,000 for Arts and Entertainment (A&E) Network's *Biography* and Lifetime's *Intimate Portrait* to $150,000 or less for most of the remaining programs, including such broadly diverse fare as VH1's *Behind the Music* and C-SPAN's *American Presidents*.[61] The forerunner and acknowledged prototype is *Biography* which was inaugurated in 1987. "The series has nearly 500 producers, associate producers, researchers and technical staff members deployed in up to 10 different production teams around the country churning out 130 episodes a year."[62] In addition, the index of historical (and contemporary) individuals and couples featured on *Biography*—from Thomas Jefferson to Jackie Robinson to Pocahontas and John Smith to Abraham and Mary Todd Lincoln—is sweeping and diverse.

All of Burns's biographical documentaries typically take years to produce. "An episode of *Biography*," in contrast, "is created in less than six months, and often in a matter of weeks." *Biography*'s executive producer Michael Cascio explains that "[w]e're trying to develop a style without having to linger on a meadow for 45 seconds. *Biography* speaks the language of TV. We don't try to pretend that we're doing arty independent cinema."[63] Despite such an obvious reference to public television, Cascio's producers do adapt highly derivative stylistics, which are often reminiscent of PBS nonfiction à la Ken Burns, as well as a pastiche of other conventions borrowed from TV news and prime-time dramatic storytelling. Still, the impact and popularity of A&E's *Biography* is undeniable. The program now averages a nightly viewership of nearly 3 million, spawning videotapes, CDs, a magazine called *Biography* with a 2 million readership, and a newly launched all-biography channel.[64] *Biography* is a representative example of how history is often framed in highly standardized and melodramatic ways on TV, mainly to be marketed and sold directly to American consumers as a commodity.

The distribution of Ken Burns's historical biographies also reflects his increasing awareness of marketplace realities. One of the more obvious offshoots of his relationship with General Motors is the literal transformation of each of these two-part biographies into a finely coordinated and expansive multimedia project. As with *Thomas Jefferson,* the four *American Lives* episodes so far have each had a running time between three and four hours, being first introduced to the public as an abbreviated miniseries over two successive evenings. The special event status of these shorter works are further heightened and confirmed by the launching of their own state-of-the-art websites, extensive press kits and promotional brochures, tightly scheduled press junkets for Burns and his coproducers, 800 numbers to facilitate the aggressive promotions conducted by PBS Video and the publishers of the companion book titles, and the subsequent distribution and syndication of the programs across an array of domestic and overseas television, cable, and home video outlets. Although Florentine Films/American Documentaries is still technically an independent production company, its partnership with General Motors ensures a cutting-edge level of marketing and distribution that approximates the commercial television sector. Burns's success in this regard, moreover, has profoundly altered the way other high-profile documentaries are promoted and released to their audiences by PBS and cable networks, such as the History Channel and TNT (Turner Network Television), using tactics once reserved only for feature films and major prime-time series.

Ken Burns and his colleagues have additionally innovated newer ways of raising money beyond the usual solicitation of corporate underwriters, philanthropic foundations, and public granting agencies. Florentine Films/American Documentaries has recently acquired substantial support from various state tourism offices. These production grants have consistently covered from one-quarter to one-third of the budgets for the five biographies in the *American Lives* film project. Virginia tourism officials actually first approached Burns about a partnership concerning *Thomas Jefferson* after noting that visitors to the state's "battlefields and landmarks jumped 50 percent" following the telecasting of *The Civil War*. Burns was surprised by the overture and welcomed the opportunity: "No time in 25 years of film making had anyone come to me and ask, do you want money to make a film."[65] The Virginia legislature allocated "a $350,000 grant toward the making of Burns' $1.3 million project."[66] The state then marshaled a coordinated marketing campaign called "Jefferson's Virginia," publicizing nine historic sites, which dovetailed strategically with the public television debut and rebroadcasts of *Thomas Jefferson*.[67] In exchange, brief promotional spots, including a toll-free telephone number, were shown at the beginning and the end of the two parts of the Jefferson biography, and promotional brochures were placed in the packaging of the videotaped version. During the following year, tourism increased 40 percent at the nine "Jefferson's Virginia" locations, while state representatives reaped an estimated four-to-one return on their initial investment.[68]

Based on this model, Montana contributed a similar proportion to *Lewis & Clark: The Journey of the Corps of Discovery;* Illinois likewise invested in *Frank Lloyd Wright;* and New York shared in the cost of *Not For Ourselves Alone: The Story of Elizabeth Cady Stanton and Susan B. Anthony*. The timing of these joint ventures turned out to be serendipitous since congressional budgetary cuts in the mid-1990s forced the National Endowment for the Humanities to slash their two decades old funding support for public programs by nearly 60 percent. NEH had been "the primary source of financing for projects that [brought] together film makers and academic historians."[69] This shift away from relying so heavily on NEH backing to such open entrepreneurial collaborations with state tourism departments resulted in an expected rise in criticism from both inside and outside the academy. Considerable alarm among Mark Twain aficionados, for example, followed closely on the heals of the December 1998 announcement that Burns was being awarded a $530,000 grant from Connecticut to help underwrite his upcoming film on the author.[70] A representative reprimand of the filmmaker appeared immediately in a *Mark Twain Newsletter* editorial: "Is Ken Burns's much anticipated documentary being sold to the highest bidder?"[71] The worry expressed in this and similar critiques is whether or not Burns and his colleagues would now be more likely to slant their biographies in favor of the states that sponsor them. *Mark Twain* coproducer Dayton Duncan answers such allegations directly:

> We're very grateful to Connecticut for their production grant—which amounts to less than a third of our funding. But we never considered nor promised "slanting" our story in one direction or another. That's not the way

we make our documentaries, and I think our past films and our reputation prove it. Long before we had any funding for this project, we were excited about the Twain House in Hartford, which is so well preserved and such a wonderful window into important aspects of his life: the happiest moments of his family and one of his saddest moments (the death of his eldest daughter, Susy), the place that symbolizes both the pinnacle of his finances and the depths of his bankruptcy. Just as "Thomas Jefferson" used Monticello as a crucial visual setting for exploring Jefferson, we've always thought that the house on Farmington Avenue would provide a magnificent backdrop for exploring a crucial segment of Mark Twain's life.[72]

The biographies in *American Lives* still subscribe to the American-original prototype, but they now have developed to a point where a more fundamental underlying principle guides their common approaches to plotting and character. Burns discovered a small book by Robert Penn Warren entitled *The Legacy of the Civil War* while researching *The Civil War* in the mid-1980s: "I stumbled across an interesting passage," the producer-director remembers, "a 'voice' if you will, that led me ultimately to Thomas Jefferson . . . [Warren wrote that] 'all the self-divisions of conflicts within individuals become a series of mirrors in which the plight of the country is reflected.'"[73] *Thomas Jefferson* certainly underscores the glaring contradictions inside the author of the Declaration of Independence who was also an extensive slave owner his entire adult life. Taking his lead from Robert Penn Warren, Burns's attitude toward Jefferson grew into an "accept[ance of] his self-divisions as a great mirror of our own possibilities and failings and to go forward."[74]

Similarly, Ken Burns began envisioning his other *American Lives* subjects. *Lewis & Clark,* for instance, is framed and presented as two parts of the same whole: "They were led by two utterly different men," relates narrator Hal Holbrook, "One outgoing and self-confident (Clark). The other brilliant but troubled (Lewis). Representing a nation that celebrated individual achievement, they would rely instead on cooperation and teamwork." *Not For Ourselves Alone: The Story of Elizabeth Cady Stanton and Susan B. Anthony* is similarly held together throughout by an apparent attraction of opposites. Narrator, Sally Kellerman, announces:

They could not have been more different. Stanton was born to wealth and comfort, was married, and the mother of seven children. She was witty and hospitable, fond of good food and fine clothes, but she was also an uncompromising revolutionary—"a many idea-ed woman," her daughter called her, who dared to proclaim to the world that women had the right to vote. Anthony was a brilliant strategist, willing to tack to the left or to the right, if by so doing she could steer the women's suffrage movement toward its goal. Though she never held public office, she was the nation's first great woman politician, "Aunt Susan," to a whole generation of young women. Despite their differences, the two would work together for more than half a century to better the lives of women everywhere.

Even in the single character studies, *Frank Lloyd Wright* and *Mark Twain,* Burns often portrays these historical figures as living their professional and private lives at polar extremes. In the case of the architect, for example, Meryle Secrest, one of his leading biographers, summarizes in the film's introduction that "one can look at [Frank Lloyd Wright] and be awed by the dimensions of his personality and the achievement because we are looking at something we seldom see in real life—which is a genius. On the other hand, when you look at who he was as a human being, he was so at the mercy of his emotions that you think he's at the other spectrum—he's barely a human being."[75] In entirely different ways, *Mark Twain,* too, is replete with countervailing tendencies: The filmed biography features the era's greatest living humorist who was also given over to severe bouts of melancholia and pessimism; he is characterized as courting his own celebrity and cultivating an extravagant lifestyle while also thinking nothing of satirizing those who lived in an analogous fashion. Like Jefferson before them, Wright and Twain are utterly American in their self-divided temperaments and contradictory actions. As architectural historian Vincent Scully recounts in Part II of *Frank Lloyd Wright:* "I think of [Wright] always as the great American confidence man who is changing the world according to his own images, who is wearing many disguises all the time. He's right out of Mark Twain."

The American originals in Ken Burns's historical biographies, therefore, have many distinctive faces and facades, but they all are similar in that they struggle with their own mirrored reflections which either lie buried deep within their own psyches, as in *Thomas Jefferson, Frank Lloyd Wright,* and *Mark Twain,* or are made apparent in the presence of an indispensable partner whose attributes and talents somehow complete his or her opposite number, as in *Lewis & Clark* and *Elizabeth Cady Stanton and Susan B. Anthony.* The foremost motivation behind all of Ken Burns's biographical plot lines is *e pluribus unum* which accurately expresses the producer-director's liberal pluralist orientation. Each and every *American Lives* episode, in other words, is a narrative of integration which either embraces all the nagging self-divisions within one subject's character or consolidates two constituent though dissimilar individuals into an historic partnership. These integration narratives, moreover, reflect symbolically on the nation as a whole. As historian Stephen E. Ambrose concludes toward the end of *Lewis & Clark,* for instance, "Lewis and Clark . . . provide us with a sense of national unity that transcends time and distance and place and brings us together from coast to coast."

In his *American Lives* series, Ken Burns is, furthermore, operating within the traditional confines of the biographical genre which dates back to the "ancient art of telling life stories."[76] For film and television biographies, in particular, the conventional approach to this story form is "divided by tone with the hagiographic bio-pic offering a model of achievement . . . at one extreme, and a satire . . . with its attitude of aggressive debunking at the other."[77] What Burns and his colleagues are able to accomplish by depicting their biographical figures in antithetical terms (e.g., visionary/hypocrite, self-confident/troubled, genius/subhuman, theoretical/pragmatic, humorous/misanthropic) is the merger of the inspirational nature of older-fashioned hagiographies with the more contemporary tendency to doubt and unmask those historical men and women who in hindsight look too good to be true. Usually personal shortcomings qualify the public brilliance and accomplishments in *American Lives,* such as the stark inconsistencies between Jefferson's words and his actions on race, Lewis's "depressions of mind" and his eventual suicide, Stanton's intemperate remarks about African-Americans during a disappointing political setback, and Twain's intermittent darker moments. Still, *American Lives* are ultimately fables of reconciliation and redemption. Even the most unlikable and seemingly irredeemable character in the series is accepted and vindicated at the end of his biography by the sheer volume and scale of his contributions to American architecture and culture and his unflagging energy and passion for living. "Frank Lloyd Wright broke all the rules in his art and in his life," describes Edward Herrmann, the narrator, "He was controversial, notorious, utterly unpredictable. He boasted of his genius with an arrogance and bombast that outraged his enemies and bewildered his friends, risked his career in a series of scandalous affairs, and suffered terrible personal tragedy, but through it all never stopped dreaming of new ways to build. He was, one of his draftsman remembered, '200 percent alive.'"

Following the precedent set by *Thomas Jefferson,* all of the *American Lives* episodes are, finally, about using the past to make better sense of the present and to explore future options. "History, it seems, to me," explains Burns, "is really not about the past; it's about the present. We define ourselves now by the subjects we choose from the past and the way each succeeding generation interprets those subjects. They are more a mirror of how we are now than they are a literal guide of what went before."[78] Just as *Thomas Jefferson* evoked an assortment of present-day issues and concerns, *Lewis & Clark*'s Corps of Discovery serves as a newly relevant metaphor for a diverse and disjointed collection of individuals at the start of their trip (i.e., young army men from Kentucky and New Hampshire; French Canadian boatmen; an African American slave; a Shoshone woman and her infant son), venturing together into uncharted territory and encountering an uncertain multicultural future (i.e., establishing a series of relations with the Lakotas, Mandans, Hidatsas, Shosones, Clatsops, and Nez Percé, among other native American tribes). Later on, *Lewis & Clark* develops into an inspirational adventure story where the explorers only succeed by eventually coming together and realizing themselves as the Corps of Discovery (i.e., *e pluribus unum*) despite being pushed to their limits in the process and suffering internal conflicts and casualties during the journey (and the death of Meriwether Lewis as a lingering aftereffect). Over 20 million viewers tuned into this contemporary updating of the *Lewis & Clark* narrative on November 4 and 5, 1997, making this reconciliatory and redemptive fable about the historic advantages of diversity during the early days of the republic the third most watched program ever on PBS after Burns's own *The Civil War* and *Baseball.*[79]

The Lewis and Clark expedition "begins . . . in Jefferson's mind," claims coproducer Dayton Duncan, "and his hand is omnipresent through it."[80] Jefferson, America's first distinguished (albeit amateur) architect, also steers Ken Burns and his colleagues toward an investigation into the life and artistic accomplishments of *Frank Lloyd Wright,* the nation's greatest architect ever. "[T]he fault line that our film exists on," notes the producer-director, "is that tension between the personal and the public."[81] This biography utilizes the intimate aesthetics of television by providing a close and revealing look at one of the century's towering creative figures who often indulged in overly egotistical and preening behavior. "From the very beginning he conceived of himself as a public figure," explains coproducer Lynn Novick, "and he knew he had to play a part to get clients and to get attention. And he consciously affected a cape, cane, certain kind of hat and a posture . . . whenever he walked into a room, heads turned even though he was only 5-foot-6!"[82] *Frank Lloyd Wright* uses its subject to underscore the rising national fixation on celebrity, scandal, and controversy in the coming media age. This production, moreover, illustrates the power of Wright's "organic" vision by employing first-person camera techniques to literally transport the viewer inside many of his more innovative structural designs. "This is a guy who lives the American century," adds Burns, "he is the prototypical American in every way."[83]

The two protagonists in *Not For Ourselves Alone: The Story of Elizabeth Cady Stanton and Susan B. Anthony* are similarly paradigmatic in their achievements and contributions to the United States, even if this seminal chapter in the country's history is not as well known as the other narrative accounts in the *American Lives* series. Chronicling their importance for a nationwide audience during prime time, in fact, provided an added sense of mission and discovery for the filmmakers. "I really hope that viewers will realize after seeing this film that here are two women who have been unjustly neglected in American history," maintains coproducer Paul Barnes.[84] Stanton and Anthony pioneered and gave voice to the women's movement for the three-quarters of a century before the passage of the Nineteenth Amendment in 1920, which finally accorded women the right to vote. Although suffrage is probably their most important legacy, attained more than a decade after each had died, their brilliance and tenacity also led to advances in education, divorce law, and the right to own property for women, among many other issues. Their combined efforts genuinely transformed the way that more than half the population was perceived and treated in American society. *Not For Ourselves Alone,* furthermore, eludes one of the inherent challenges in the biographical genre, which is finding novel ways to build dramatic tension in a plot line that is generally familiar to most people. The unintentional benefit of recovering the comparatively hidden history of Stanton and Anthony is to render the *American Lives* formula that much more affecting for the estimated 14 million viewers who watched this dual biography on November 7 and 8, 1999.

Not For Ourselves Alone emphasizes the close connection between the women's rights and abolitionist movements in the decades leading up to the Civil War. *Mark Twain* too links the enduring significance of this author to his subtle depictions of race relations in the pre-Civil War south in his profoundly influential writings, especially *The Adventures of Huckleberry Finn* (1884). Ken Burns has always insisted that his continuing examination of America's racial heritage is the "connecting thread" in all of his work.[85] This latest biography once again highlights his abiding interest in this topic as Twain is observed cautioning his fellow citizens: "This is how you are, like it or not." The relevance of *Mark Twain* is evident too in its emphasis on the writer's anti-imperialistic stance during an era when the United States was first assuming its current position as a world power, as well as the way that "the most conspicuous human being on earth" presaged the celebrity culture. As with all the *American Lives* episodes, *Mark Twain* may in the end be mistaken for a full and exhaustive account of the author and his times when, in fact, this television biography as popular history is far more concerned with exploring the pertinence of Twain's life and legacy in the present than in rendering a complete and definitive record of what actually happened in the past.

In choosing to create historical documentaries in a shorter biographical format, Ken Burns and his colleagues have assembled a series of tightly structured life stories on film whose integration narratives literally enact the liberal pluralist promise of America. In so doing, the filmmakers mediate between the hagiographic and satiric traditions of the movie biography by portraying an emerging roster of self-divided individuals and couples who all mirror a wide assortment of admirable and defective qualities in the national character which finally are resolved for the better in these five fables of reconciliation and redemption. *American Lives* as popular history is personal in its biographical approach, consensual and majoritarian in its ideological outlook, and present-minded in its attitude toward the past. These television specials are also consistent ratings winners for PBS, marketing innovators for the documentary field in general, and cost-effective promotions for General Motors (for whom "Burns's projects amount to a small portion of its overall advertising budget").[86] Florentine Films/American Documentaries' heightened commercial profile is still an undeniable consequence of its partnership with GM, in which even its shorter format films, such as the episodes of *American Lives,* are afforded major production-distribution and educational outreach support far in excess of the typical levels available anywhere else in public TV or in the world of independent nonfiction filmmaking.

Upon the completion and premiere of *Baseball* in September 1994, Burns concurrently produced and directed "more or less full time, despite the other films that we work full time on," a 10-part, 19-hour, epic-sized series entitled *Jazz.* Burns observed, "I see this as a final third in a trilogy of American life that began with *The Civil War,* continued with our film, *Baseball,* and will sort of come to its fruition in this history of jazz. The Civil War defined us. Baseball told us what we had become from that defining event, and jazz is a way to point to the possibilities of our country, the redemptive soul of America in this most original music that we, Americans, have invented."[87]

During the six years that Burns and his creative team went back and forth between *American Lives* and *Jazz,* this concluding chapter of his trilogy evolved into Florentine Films/American Documentaries' largest and most ambitious project to date. Burns's coproducer on *Jazz* was once again Lynn Novick, reprising the professional duties and responsibilities she first assumed on *Baseball* and fully realized on *Frank Lloyd Wright.* "Ken really knows what he wants and where he wants the film to go," declares Novick. "It is quite inspiring to see that focus and direction able to marshal all these forces and kind of make everything come together. So I've had a great teacher."[88] *Frank Lloyd Wright* was, moreover, for Burns, the "first film in which I've shared equally the credit and deservedly so. Lynn and I have worked together for a long, long time, so sometimes even the question of the division of labor is besides the point."[89]

Ken Burns further reveals that "I look at any of the [films] that I've done as efforts at trying to be better, efforts of trying to learn something new, of improving a facet of myself, educating [a] part that needed growing up, and I don't know in one's whole life long whether you can ever stop doing that."[90] And, he says, "a lot of [*Frank Lloyd Wright*] had to do with sharing . . . [t]hat is, in a funny way, as creative a movement for me as thinking up a new way to shoot something or a new way to approach a scene or a new way to structure a film."[91] The most basic challenge of *American Lives,* in the end, and the most indispensable lesson that Burns absorbed while often overseeing as many as three highly complex projects at the same time in various stages of development "was about how you share it, how you allow your vision to be incorporated by someone else and to be changed by the presence of someone else."[92] *Jazz* was the most involved and elaborate result of such intense collaboration between a close-knit group of coproducers, writers, editors, cinematographers, historical advisors and consultants, and other staff and crew members at Florentine Films/American Documentaries. Many of these associates and friends have now worked together for ten to twenty years or more. *Jazz,* in this regard, reflects the maturing nature of their professional and personal relationships as well as Ken Burns's continued growth and acuity as an executive producer. It is also the clearest indication yet of the consistency of Burns's artistic vision at midcareer.

Ken Burns's America Reconsidered: Mainstreaming *Jazz* (2001) for a National Audience

MAKING ROOM FOR THE MUSIC

Jazz is art, not sociology. It is primarily an African American creation but it belongs to all of us, and the chance to bring it alive for a national audience in serious need of being reminded of the astonishing feats Americans can perform when not held back by prejudice and parochialism, is a once-in-a-lifetime opportunity for which I shall always be grateful.

—Geoffrey C. Ward, 1996 [1]

More than half a century ago, F. Scott Fitzgerald hardly imagined the many comebacks that television would allow by its steady stream of reruns, revivals, and return engagements when he wrote his now-classic line, "there are no second acts in American lives." [2] TV is full of second acts and more; it even provides succeeding generations with ready access to the once famous but now forgotten who somehow take on a fresh new relevance with the passage of time. Geoff Ward, for one, never lost his lifelong interest in jazz, including his fascination with the music's many innovative composers and performers. He brought his small but burgeoning record collection with him at 14 when his family left

Chicago to live in India during his high school years: "[O]n some level jazz music kept me rooted at home."[3] When he journeyed again at 16 to briefly attend school in France, he developed a deep attachment for Louis Armstrong and remembers playing "West End Blues" "a thousand times over that summer."[4]

Ward was actually the first person to raise "the possibility of producing a history of jazz" on film to Ken Burns when they were both fully absorbed in the final stages of *The Civil War* in 1988.[5] By that time, the writer had been acquiring jazz records for almost 40 years. He had also amassed an extensive library of books on the subject, had published several articles in *American Heritage* and other magazines on the topic, and would play jazz as background music every day while writing. When Burns finally agreed to produce and direct *Jazz* after the release of *Baseball,* Ward wrote a letter to his old friend and colleague confiding that "[i]t does seem to me that I've been waiting for this project most of my life."[6]

The prospects of pursuing jazz as a multipart documentary were discussed on and off by various creative personnel at Florentine Films for nearly five years until a seemingly unrelated remark by Gerald Early, a professor at Washington University and one of the principal scholarly advisors on *Baseball,* propelled the subject of this musical genre front and center in 1993. During one of the last interviews Burns conducted for that miniseries, in a commentary that eventually was included in the opening episode of *Baseball,* he asserted: "I enjoy the game because it is a beautifully designed game—it's a beautiful game to watch—but principally because it makes me feel more American. It makes me feel connected to this culture, and I think there are only three things that America will be known for 2,000 years from now when they study this civilization—the Constitution, jazz music, and baseball. They're the three most beautifully designed things this country ever produced."

"That gave me an intellectual rationale to pursue a jazz film," Ken Burns recalls, but he still deferred final approval until feeling that he was fully committed to the project in a more "emotional fashion."[7] This typically intuitive resolution for Burns came while he was at home with his two daughters watching *Baseball*'s "Fifth Inning: Shadow Ball (1930-1940)" on September 22, 1994, during its initial PBS telecast. In the chapter entitled "You Lucky Bum," the filmmaker remembers:

> There was a scene about a World Series, the 1932 World Series between the Cubs and Yankees in which Babe Ruth is alleged to have called his shot. We used a particular piece of music by Count Basie called "Tickle Toe," and I remember loving editing it with the editor, Yaffa Lerea . . . and I remember that I loved working with it in the editing room and I loved seeing it then. I knew at that moment in my bones—I'd been thinking about jazz and we had been talking about jazz as an intellectual thing for many years, but it wasn't

until seeing "Tickle Toe" by Count Basie in the fifth episode of *Baseball* that I knew I had to do jazz.[8]

Ken Burns thus set the preparation phase of the Jazz Film Project into motion. Geoff Ward agreed at once to act as head scriptwriter. Burns invited his *Baseball* collaborator Lynn Novick to share producer duties with him again, as he would also direct all the creative aspects of the upcoming miniseries. Together the trio "drew up plans for an eight-episode, approximately twelve-hour history of jazz" in the fall of 1994.[9] Burns realized from the start that *Jazz* presented a whole new aesthetic challenge: Sound would now be more important than in any of his previous films, even becoming the sole focus of the narrative at certain key intervals rather than being relegated to a supporting role behind the visuals. "We're making room for the music," he noted in retrospect, "and listening to it on a level we haven't before."[10]

Preliminary funding and the remainder of the creative team fell into place over the next year. A provisional budget of $9,878,085 was drafted in early 1995 and by spring General Motors had committed $3 million in production costs and a similar amount of money for eventual advertising, promotion, and educational outreach.[11] The continuing popular success of *Baseball* boosted the rising prospects of the *Jazz* film project as a complete rerun of the nine-inning series by PBS in January 1995 increased the aggregate viewership for the two telecasts of the miniseries to nearly 45 million people. Another major vote of confidence came during the summer months as the Public Broadcasting Service awarded Florentine Films an additional $2 million for *Jazz*.[12] At the same time, Burns and Novick hired two other veteran coproducers, Madison Lacy and Peter Miller, and began assembling an advisory board of critics, scholars, and performers headed by Wynton Marsalis as senior creative consultant.[13] Burns first met Marsalis after a speaking engagement at the Metropolitan Museum of Art in 1991 where the musician suggested to the director that he consider producing a history of jazz on film. "As I told you after seeing the Civil War documentary," wrote Marsalis in a June 1995 letter accepting Burns's invitation to become the lead advisor on *Jazz*, "you have the soul to do this very difficult project in fine and high style . . . Like you," the trumpeter concedes, "I am anxious to undertake this project and am willing to do *whatever it takes* to assist your effort."[14]

Spirits were high and climbing as Burns and Novick, especially, began the extended process of learning as much as they could about jazz. "The initial thing we try to do," Novick describes, "is immerse ourselves in the subject through meetings with scholars, reading all the books we could get our hands on and so forth. And in the case of 'Jazz' . . . listening to the music."[15] Novick also opened a New York production office for the *Jazz* film project with Lacy and Miller in October 1995, where they set about the business of additional fundraising, researching sound recordings, film clips, and photographs, and consulting with various advisors, which included conducting the first two filmed interviews for the miniseries with Marsalis and Gerald Early.

Burns enlisted Buddy Squires as principal cinematographer and Paul Barnes as supervising film editor. Burns and Geoff Ward, moreover, concentrated much of their energies at this point on creating a beginning treatment of the entire *Jazz* narrative by the fall of 1996 so that they could solicit preliminary feedback from their board of advisors.[16] Ward, of course, brought his years of "quiet obsess[ion]" with the music to bear in fashioning the structural outlines of *Jazz*. "At the same time," he explains, "Ken was thinking about the whole project on his own. Then Ken and I sat down and went through it. He made a lot of wonderful suggestions, we rearranged, cut and pasted, and went at it all over again, producing two or three drafts."[17] The latest version of the working screenplay was subsequently delivered to the other producers and editors in January 1998 to begin a rough assembly of the live cinematography, archival photographs, filmed interviews, taped voices, period recordings, and new renditions of several featured songs recorded specifically for *Jazz*. "We don't do this business without collaboration," Burns summarizes, "I'm in some ways more like the conductor who comes up and has sort of this easy task, in a way, of orchestrating the extraordinary talents of many people who, in their combined efforts, make what we call a film by Ken Burns."[18]

THE KEN BURNS FILM AS BRICOLAGE

My films are really made in the editing room and feature films are not. They're made shooting. You have to previsualize everything. I am dumb. I have to work a structure and tear it apart and rework it and do that, and find the truth in it in the editing room.

—*Ken Burns, 1993*[19]

Bricoleur is the traditional French word for handyman or jack-of-all-trades. Anthropologist Claude Lévi-Strauss adapted this term during the early 1960s in such books as *The Savage Mind* to refer specifically to someone who still works with his or her hands but does so to make sense of the world in a mainly poetic (and resolutely unscientific) manner of producing analogies out of the raw materials of culture. Bricolage, in this sense, can be a highly creative and improvisational activity; according to Lévi-Strauss, mythmaking is "a kind of intellectual 'bricolage'" where the bricoleur constructs meanings "by means of a heterogeneous repertoire which, even if extensive, is nevertheless limited" to the prevailing ways of seeing and the available resources in the surrounding environment.[20] A bricoleur, moreover, "'speaks' not only *with* things, but also through the medium of things: giving an account of his personality and life by the choices he makes between the limited possibilities."[21]

Ken Burns's documentary histories are likewise built through an extremely complex and painstaking process of bricolage. The producer-director and his creative team spend years researching, retrieving, and artistically reassembling live and stock footage, photographs, prints, paintings, daguerreotypes, images of historical buildings and artifacts, narration, recorded commentaries, and music, along with a wide assortment of other media components. "At a certain point we just say 'Look,'" declares longtime collaborator Lynn Novick, "we have enough material to make a good film. And even though there are other people out there who have great stories to tell, we simply have to say that we have a limited budget and time, and we have to take what we've got and make that work."[22] Burns and his associates then generate a series of poetic analogies, both large and small. *The Civil War*, in this way, becomes America's *Iliad*, while *Baseball* is its *Odyssey*. Thomas Jefferson is similarly "a kind of Rosetta Stone of the American experience"; Monticello, in a more individualized context, stands as "a metaphor for Jefferson's soul."[23] These analogies are not matters of right or wrong per se; they are mythic strategies devised by the filmmaker-as-bricoleur for determining the relevance of certain people, places, and events from the nation's history for himself, his creative team, and a new generation of American viewers.

Jazz continues this strategy. Ken Burns and his colleagues worked on this project for the better part of six years. They fashioned this nearly 19-hour miniseries out of 497 separate pieces of music, 2,400 stills, 2,000 motion picture clips, and portions of 75 filmed interviews.[24] Their guiding consideration, as before, was a slightly adapted version of Burns's original query—"Who are we as Americans?" In subsequent grant applications, Novick, Lacy, and Miller articulated that same overriding spirit of search and discovery that has motivated all of Burns's projects at Florentine Films/American Documentaries: "As documentary filmmakers engaged for the past twenty years with exploring what the history of our nation can tell us about ourselves, we have been concerned above all with trying to answer one simple question: 'What does an investigation of the past tell us about who we are and what we have become?'"[25] This inquiry— as do all seminal questions—guides and influences the sources they consult and the answers they ultimately find.

Jazz became "the sound track of America" for Burns, according to his own description, and "a very accurate prism through which you could see refracted tendencies of my country. I found what I wanted . . . [o]n every level—mythological, sensual, political, and social—it informed me about who we are as a people. I didn't expect to learn so much."[26] Jazz in all of its richness as a century-old artistic and cultural form thus slowly though inexorably emerges in the producer-director's latest miniseries as yet another in a long line of historical analogies that points Burns in the direction of finding and expressing his vision of America, an outlook he has been creating and actively fine tuning with the help of his collaborators for more than 30 years. Burns, the bricoleur, reassembles anew each time he makes a film a patchwork quilt of Americana that is inevitably a mixture of history and myth. Such is his method of reproducing the past on television. "As a filmmaker and amateur historian," he concludes, "I am aiming at a wide general audience, and what I want to do is give those Americans who have perhaps forgotten about the music that is the soul of their country a chance to reexamine it. I don't wish this to be the definitive last statement on jazz, but a kind of way for Americans to be reintroduced to it."[27]

AMERICAN LIVES WRIT LARGE

> Memory takes a lot of poetic license. It omits some details; others are
> exaggerated, according to the emotional value of the articles it touches, for
> memory is seated predominantly in the heart.
>
> —*Tennessee Williams*, The Glass Menagerie,
> *stage directions for Act One, Scene One*[28]

The elongated production period for *Jazz* overlapped with the making of all five *American Lives* installments. The premiere episode, *Thomas Jefferson,* was still being scripted and shot when the *Jazz* film project was launched in the fall of 1994, and *Mark Twain,* the latest of these two-part character studies in what will surely become a continuing series, was just beginning its initial edit preparation as the final video transfers of *Jazz* were being finalized in July 2000. There are, in turn, striking similarities between the biographical and ideological parameters of these five shorter historical documentaries and the story and thematic concerns of Burns's much longer miniseries on this most indigenous of musical genres. *Jazz* essentially extends the design and outlook of *American Lives* in its 19-hour integration narrative and its abiding faith in the reconciliatory and redemptive power of the music. The opening 5-minute introduction of "Episode One: 'Gumbo' (Beginnings to 1917)" establishes this clear affinity from the outset.

Jazz starts out with a shimmering black-and-white bird's-eye view photograph of Manhattan at night as the sounds of the city effortlessly dissolve into Louis Armstrong and his orchestra playing "Star Dust." The camera slowly tilts uptown for 27 seconds before it cuts to a medium shot of Armstrong blowing his trumpet on a bandstand with his eyes characteristically closed. This second shot of *Jazz*'s lead character lasts almost as long as the previous one, as Wynton Marsalis delivers his opening remarks in a voice-over: "Jazz music objectifies America. It's an art form that can give us a painless way of understanding ourselves." Within the first minute of the series, the centrality of Armstrong is plainly foreshadowed, but even more importantly, the significance of jazz is framed as something much more than just a musical category. Marsalis appears again onscreen and his commentary continues for one full minute as he further intimates that jazz is also a kind of national symbol through which viewers can garner insights about their country and themselves.

Ken Burns agrees completely with Marsalis's utilization of jazz as a metaphor; that is why he features him so prominently in the first two minutes of the film. Burns's own employment of jazz as a historical analogy is fully consistent with the way that he approaches all of his films as bricolage. In a recent interview he elaborates that

> you have to understand that jazz is merely a delivery vehicle. So that you can
> also say that it is a film about race and race relations and prejudice. It is a film
> about wars, two world wars. It is a film about a Great Depression. It is a film
> about sex and the relationship between men and women. It is a film about
> drugs and their terrible cost and price. It is a film about the growth and decay
> of cities. But at its heart, this is a film about creativity, and specifically American
> creativity. So jazz really becomes a vessel into which one can pour questions
> about who we are in many areas that stretch beyond jazz.[29]

Burns next shifts the focus of *Jazz*'s introductory scene to the narrator, Keith David, whose rich and expressive baritone is heard offscreen over an energetic two-minute montage composed mainly of flashing neon signs, hyperkinetic lindy-hoppers, and a series of brief cameo shots of more than a dozen greats from the pantheon of jazz, including Art Tatum, Benny Goodman, Dave Brubeck, Count Basie, Billie Holiday, Charlie Parker, Dizzy Gillespie, Louis Armstrong once more, John Coltrane, and Miles Davis, among other luminaries, before this middle section of the opening scene ends with a joyous Duke Ellington playing his piano. All of these many visuals have come streaming by as an electrifying rendition of Ellington's "Take the 'A' Train" pulsates on the sound track. Almost as a soloist, David enters and exits above the music during this montage, orienting the audience with his opening salvo from Geoff Ward's poetic narration: "It is America's music. Born out of a million American negotiations: between having and not having; between happy and sad; country and city; between black and white and men and women; between the old Africa and the old Europe—which could only have happened in an entirely new world . . . 'Jazz,' drummer Art Blakey liked to say, 'washes away the dust of everyday life.' Above all, it swings."

As in *American Lives* and his other major miniseries, Burns and his creative team proffer a liberal pluralist panorama of the country which literally captures in *Jazz* the coming together of a wide variety of polar extremes based on money and possessions ("having and not having"), temperament ("happy and sad"), geography ("country and city"), race ("black and white"), gender ("men and women"), and, finally, place of origin ("old Africa and old Europe"). These binary opposites reflect more the rhetoric of myth than scholarly historical discourse and thus immediately provide a style of poetic language which is intended for filmmakers and viewers alike to rediscover and renew the common elements of their shared cultural heritage.[30] The "Gumbo" title of episode one is yet another metaphor alluding to all the musical ingredients—minstrel songs, Caribbean rhythms, marching tunes, Italian opera, ragtime syncopations, and the soulful feeling of the blues—that slowly simmer together for decades in the proverbial stew that eventually becomes jazz, another analogy representing *e pluribus unum*. New Orleans, too, is the cosmopolitan melting pot that gives birth to this wholly new and peculiarly American art form. The opening scene appropriately ends with three final impassioned observations which reflect Burns's own unabashed enthusiasm for the subject—Wynton Marsalis: "Jazz music celebrates life—human life."; Gary Giddins: "It's the ultimate in rugged individualism."; and Albert Murray: "It's the creative process incarnate."

Jazz music has rarely been embraced with such eagerness and high spirits by anyone associated with television before. The historical relationship between jazz and TV has indeed been lukewarm at best, although consulting producer Madison Lacy remembers "hav[ing] seen at least a half dozen sincere and hard fought attempts by stations and independent producers to develop such a jazz series."[31] Just as network television was emerging in the mid-to-late 1940s, jazz was on its way to becoming more of an acquired taste. By the early 1950s, it was no longer the popular music of America as it had been during its heights in the swing era just a decade earlier. Television, instead, grew up with rock 'n roll; and except for the occasional prime-time special and the infrequent guest appearances on various talk shows over the years, jazz composers and performers were largely absent from the small screen.[32]

Midway through "Episode Nine: 'The Adventure' (1956 to 1960)," Burns actually includes a 4-minute 55-second scene of the most remembered jazz program in television history. On December 6, 1957, critics Whitney Balliett and Nat Hentoff organized a live special on CBS that began with a 50-second introduction of the "all-star assemblage," according to narrator Keith David, which included "Jo Jones and Count Basie, Thelonius Monk and Coleman Hawkins, Gerry Mulligan and Ben Webster, Lester Young and Billie Holiday." The final 4 minutes and 5 seconds of the scene is devoted entirely to the performance of Holiday's own bluesy "Fine and Mellow" and the quiet drama that ensued between her and Lester Young that celebrated Friday evening.

First onscreen, then off, Hentoff relates how the one-time close friends and musical collaborators had grown estranged and drifted apart years before. Holiday is shown sitting on a stool in the spotlight, singing her song, accompanied by Ben Webster on tenor saxophone, as "Lester got up and played the purist blues I had ever heard," acknowledges Hentoff. Burns significantly pauses for a generous 22 seconds—just as he often does when rephotographing still images—to allow the viewer ample time to listen carefully and absorb the music being featured on the sound track in between the spoken commentaries. Hentoff continues offscreen as Holiday and Young appear in full view, courtesy of an original kinescope recording: "And their eyes were interlocked and she was sort of nodding and half-smiling. It was as if they were both remembering what had been, whatever that was" (a brief musical interlude cuts in for 8 seconds) "and in the control room we were all crying." "Fine and Mellow" resumes uninterrupted for another 70 seconds, focusing primary attention on the song once again, as the heartfelt reunion culminates with Keith David's understated resolution that "when the show was over, they went their separate ways."

Ken Burns had the requisite stature and track record to garner the necessary support for the *Jazz* film project where others did not. *Jazz* ended up costing $14 million, stretching seven hours more than originally planned, and expanding from eight to ten episodes in the process.[33] The first and last installments cover the most ground, with "Gumbo" beginning somewhere around 1840 with some early references to minstrelsy and then traveling forward 77 years to 1917; "Episode Ten: 'A Masterpiece by Midnight' (1960 to the Present)" surveys the final 40 years in a narrative decision that probably attracted more criticism for the series than any other from hard-core jazz aficionados.[34] "One reason we didn't do more on the current era," explains Geoff Ward, "is that at some point writing about the music becomes more journalism than history."[35] Two additional motives are also suggested in a passage from Ward's companion volume, *Jazz: A History of America's Music:* "No book, no shelf of books, could adequately map the course of jazz after 1960, let alone trace the meandering paths of all its proliferating tributaries. No Great Man can be said to have towered over everyone else, as Louis Armstrong and Charlie Parker could be said to have done in their time, but John Coltrane and Miles Davis were surely among the most influential of all post-bebop musicians, and their careers touched upon many of the most important developments in the music, both good and ill."[36]

The eight middle episodes, in contrast to the introductory and the concluding ones, encompass a total of 43 years with none spanning more than a decade (for example, "Episode Eight: 'Risk,' 1945 to 1955,") and one ("Episode Six: 'Swing: The Velocity of Celebration,' 1937 to 1939") even devoting its entire 1-hour and 42-minute running time to only two admittedly jam-packed years. The majority of attention in *Jazz* is paid to the music and its close connection to American culture from the dawning of the jazz age during the last two years of World War I through the escalating turbulence of the early 1960s. The plot line is thus organized chronologically, as are all of Burns's histories. From a purely structural point of view, however, *Jazz* is even more intrinsically biographical in its historical approach than either *The Civil War* or *Baseball.*

The narrative spine of *The Civil War* begins with the causes of the conflict, increasing momentum incrementally over nine successive episodes, before reaching its denouement with the assassination of President Lincoln and the cessation of hostilities in 1865. *Baseball*'s storyline is far less clear-cut in comparison: it is ostensibly cyclical in design with the game recurring each summer, with a brand new season year after year, and surveys approximately 75 important figures while supplying another two dozen more detailed career vignettes for those select number of superstars, managers, and team executives who made their mark on the game in some extraordinary way. *Baseball,* as a result, is filled with stops and starts, ending up far less linear and more episodic than either *The Civil War* or *Jazz.* Its most well developed subplot is actually the reappearing background story about race relations, which eventually surfaces in the 1940s and 1950s with the coming of Jackie Robinson. As significant as Robinson is to the long history of *Baseball,* though, his life story still doesn't provide the kind of structural consistency throughout the whole miniseries that Louis Armstrong's does in *Jazz,* although Armstrong, too, is ably supported in this regard by Duke Ellington and Charlie Parker.

Jazz places African Americans at the very heart of the American experience. "I contend that the Negro is the creative voice of America," insists Duke

Ellington in "Episode Seven: 'Dedicated to Chaos' (1940 to 1945)," and no two individuals figure more prominently throughout the nearly 19-hour miniseries than Louis Armstrong and Ellington. Armstrong is undeniably the star, which is highly ironic considering that all of his feature film roles when he was alive were either musical or novelty cameo appearances in such pictures as *Pennies from Heaven* (1936), *Cabin in the Sky* (1943), *High Society* (1957), and *Hello Dolly!* (1969). Louis Armstrong gets his due in *Jazz,* nevertheless, as he eclipses everyone else in the miniseries as the heroic personification of the music itself—joyous, warm, generous, creative, and improvisational. In the final minutes of "Episode Two: 'The Gift' (1917 to 1924)," for example, jazz historian and long-time WKCR-FM (Columbia University) radio personality Phil Schaap recounts: "Louis Armstrong's arrival in September of '24 is pivotal because he is the most important jazz musician on the face of the earth and he's coming to the country's biggest city, playing with its most important band [i.e., Fletcher Henderson's] . . . Jazz arrives because Louis came to New York and taught the world to swing."

Jazz's ten episodes are divided into 96 chapters, most of which are identifiable at the beginning by a white title superimposed on a black background, signaling the narrative or thematic focus of the section. At the end of "Episode Three: 'Our Language' (1924 to 1929)," for instance, two interrelated scenes comprise an 11-minute 22-second chapter entitled "Modern Time." The first scene lasts 4 minutes and 25 seconds and essentially recounts how Armstrong literally revolutionized American popular music between 1924 and 1928. With the sound of Armstrong's trumpet punctuating the background, critic Gary Giddins appears onscreen alone for 53 seconds, noting that Beethoven "was a celebrated improviser" and Bach often composed "improvisationally . . . but Armstrong and jazz [came] along at the same time that there is a technology [i.e., sound recording] [to] document . . . improvisation [which] can be just as coherent, imaginative, emotionally satisfying, and durable as a written piece of music." The remaining three and a half minutes of this scene alternates three period photographs of Armstrong with a fourth still image, a group shot of him with his Hot Five band, as Giddins explains mostly in voice-over how Armstrong, first, establishes "jazz [as] a soloist art"; second, "affirms the blues tonality" of the music; and, most importantly, "Armstrong invented what for a lack of a more specific phrase we call swing. He created modern time." As if to reinforce this final point, the scene features a breathtaking and seemingly effortless stop-time solo by Armstrong, finishing with a startling crescendo on high C, which holds for 14 seconds as the camera slowly pans another photograph of him holding out his gleaming trumpet and smiling admiringly at it.

The climax of this chapter occurs in its second and final scene which focuses solely on Armstrong's remarkably expansive and legendary interpretation of "West End Blues," a song written by King Oliver, his early mentor. A 3-minute and 40-second buildup begins with a strikingly handsome publicity shot of a young Earl Hines whose "trumpet styled piano . . . spurred [Armstrong] on to greater heights." "On June 28, 1928, Louis Armstrong and Earl Hines went into the studio," declares Keith David offscreen as the camera pans slowly to the left across a photograph of the entire band before stopping and cutting to a close-up of Armstrong; "it would become one of the best-known recordings in the history of jazz, a perfect reflection of the country in the moments before the Great Depression." The lens comes to rest on a tight shot of the words "West End Blues" and "Louis Armstrong and his Hot Five" printed around a still lushly colored red and gold OKeh 78 r.p.m. label which starts to revolve on a vintage Victrola. For the next 3 minutes and 16 seconds, 39 archival film clips from daily life—blacks and whites, men, women, and children, all engaged in ordinary activities—waft along in time to the music as Armstrong's bright and buoyant horn and his sweeter-than-usual gravely voice take center stage. Burns again shifts primary attention away from the screen to the sound track, placing his documentary montage in the full service of the music's tempo and mood, as the audience is free to contemplate without undue distraction the narrator's final assertion that West End Blues "would once and for all establish Louis Armstrong as the first great solo genius of the music."

Jazz, therefore, is grounded much more heavily in the great-man theory of history than are the first two thirds of Ken Burns's American trilogy. Despite the large number of players covered in *Baseball,* Burns and Ward employ the always fluctuating fortunes of the Boston Red Sox and the Brooklyn/Los Angeles Dodgers as their most obvious structuring device throughout the over 18 hours of that storyline; it is the one narrative strategy that keeps returning over and over again. *Jazz* likewise mentions roughly 75 important individuals, and addresses more than two dozen other characters in enough depth to accord them the main focus of at least one chapter (beginning with Buddy Bolden and continuing through Jelly Roll Morton, Sidney Bechet, Louis Armstrong, Duke Ellington, Paul Whiteman, Fletcher Henderson, Bessie Smith, Bix Beiderbecke, Fats Waller, Art Tatum, Benny Goodman, Artie Shaw, Billie Holiday, Count Basie, Coleman Hawkins, Lester Young, Chick Webb, John Hammond, Ella Fitzgerald, Charlie Parker, Dizzy Gillespie, Thelonius Monk, Dave Brubeck, Miles Davis, and John Coltrane). Still *Jazz* is far more selective and less encyclopedic than *Baseball;* it is also more disciplined in its plot construction.

Louis Armstrong, overall, appears in 19 of the 96 chapters or nearly 20 percent of the plot structure. Jazz and he are both born in the richly multi-ethnic streets of New Orleans at the turn of the twentieth century. They journey north to Chicago together during the first great migration of African Americans, sustaining and enriching each other's improbable rise to success and respectability; and Louis Armstrong and jazz eventually take New York by storm, jointly becoming household names and transforming American music and culture with their ability to swing. The whole narrative arc of *Jazz* is, thus, dependent on the life story of Armstrong and, to a lesser degree, Duke Ellington and Charlie Parker, the three men around whom Burns and his creative team anchor their nearly 19-hour plot structure as the three greatest men of jazz. Once they had established these emblematic figures as their guideposts throughout the narrative, explains Lynn Novick, "it was open season. And there are 50 intersecting parallel stories that you can be touching on at any given time. It is an incredibly, overwhelmingly complex and sophisticated story and we had an infinite number of choices to make [and] decisions to face."[37]

By "Episode Four: 'The True Welcome' (1924 to 1929)," for example, New York is the undisputed jazz capital of the nation, serving as the mecca for musicians from all over the country. The music, like the United States itself, is still segregated at the time, as Chick Webb succeeds Fletcher Henderson as the leading big-band pioneer in Harlem. In addition, Edward Kennedy Ellington, nicknamed Duke because of his elegance and sophistication, once and for all buries the legacy of minstrelsy in the wake of his orchestra's urbane sound and the stylish complexities of his own original compositions. Orchestrated jazz emerges as the wave of the future in "Episode Five: 'Swing: Pure Pleasure' (1935 to 1937)," unleashing a national dance craze, reviving a moribund recording industry, and, most importantly, becoming the defining music for a whole new generation of Americans.

The next two episodes build on the already well established mood of expectant excitement and innovation in jazz, as white bandleaders such as Benny Goodman, Tommy Dorsey, and Artie Shaw reflect the influences of their black predecessors in adapting and popularizing swing for audiences from coast to coast. Seminal moments abound everywhere: Benny Goodman's concert at the Palomar Ballroom in Los Angeles during the summer of 1935 eventually earns him the title "King of Swing"; Goodman's trip uptown to the Savoy Ballroom in Harlem two years later results in a legendary battle of the bands with Chick Webb who furiously outswings his visitor; Count Basie's arrival (Episode Six) with a bluesier Kansas City sound reignites jazz by returning it to its roots; the appearance of Benny Goodman's orchestra at Carnegie Hall in January 1938 further legitimizes the new music; and, later that spring, the first outdoor jazz festival, headlined by the Count Basie Orchestra, attracts 24,000 adult and teenage fans to Randalls Island in New York. Most of Episode Seven covers the peak years of swing during World War II, along with the emergence of Charlie "Bird" Parker who brilliantly revolutionizes the music, once again, with a new style of playing eventually called bebop. Parker influences yet another generation of postwar musicians with his astonishing virtuosity as well as with an ever hipper, cooler, and more forbidding lifestyle.

After 1960, though, *Jazz*'s storyline loses much of its momentum, reflecting the great-man priorities of the storytelling as well as the liberal pluralist orientation of Burns and his closest collaborators. Through all the changes, Louis Armstrong and Duke Ellington continue to appear at select intervals, still touring and expanding their musical repertoires, providing living examples of jazz's enduring vitality. When Charlie Parker self-destructs at age 34 toward the end of Episode Eight, however, much of the joy and exuberance also evaporates from the miniseries. Burns attempts to find some kind of redemptive antidote to Parker's tragic demise by adding a five-minute coda to the episode, recounting how Miles Davis kicked his own heroin habit. "All I could think of was playing music," Davis is quoted as saying, "and making up for all the time I had lost." Despite this temporary recovery, however, *Jazz* is fast becoming a story of unrealized hopes and dreams and, as the title of Parker's final chapter suggests, "The Future Unlived."

Two more important premature deaths occur midway through Episode Nine. "By the time Lester Young died" of alcohol abuse on March 15, 1959, Keith David solemnly announces, "his old friend Billie Holiday was unrecognizable." "Perhaps the greatest of all jazz singers," she, too, expires at the young age of 44 just two months later because of drugs. Each of these mounting casualties now work as a counterweight against the sense of jazz as an "affirmation in the face of adversity" that Burns and his creative team cultivate so assiduously through the first seven and a half episodes.[38] Only 38 minutes before the end of the tenth and final episode, a live fixed shot of a modest two-story brick house in Queens appears on the screen. "In the early morning hours of July 6, 1971," reports David again, "Louis Armstrong, the most important figure in the history of jazz, died at his home." Then, a mere nine minutes later and only 29 minutes away from the end of the entire series, David reports that Duke Ellington, "considered by many the greatest of all American composers," died on May 24, 1974. A lone solitary shot of Ellington's gravestone follows, as the narrator continues: "He was buried in Woodlawn cemetery in the Bronx, not far from Louis Armstrong, and next to his mother who was the first to tell him he was 'blessed.'"

Although the remaining footage does attempt to survey some of the post-1970 "meandering paths" of jazz, as described by Geoff Ward in the companion volume to the miniseries, the logic behind this particular narrative is clearly coming to an end. There are no remaining great men left to chronicle, and Ward's many "proliferating tributaries" run counter to a view of American society and culture that is integrative and majoritarian rather than factionalized into a diverse and often bickering array of special interests. Jazz music is, above all else, the aural embodiment of *e pluribus unum* for Burns. As he explains, "*Jazz* is this wonderful portrait of not only the twentieth century, but of our redemptive future possibilities. In *Jazz*, we see the ultimate of the democratic idea. Different races, different styles, different souls, all negotiating their agendas together. When jazz works, it's a kind of model . . . of what democracy is about."[39] As the music and its storyline ceases to enact and thus symbolize this overriding metaphor, the narrative has nowhere else to go but to conclude as quickly and unobtrusively as possible, finishing with a brief survey of new jazz talent that has unexpectedly emerged and continues to grow in size and influence even today.

A CONSISTENCY OF VISION AT MIDCAREER

I'm doing this film because I'm interested in race. Race is the soul of the country, and nowhere is it more evident than in jazz, where a music came out of the black community and with great generosity was shared with the country.

—*Ken Burns, 2000*[40]

Duke Ellington's skills as a bandleader and gifts as a pianist and performer are explored in the chapter "We Need To Be Free," in Episode Seven. He is described as "a miraculous jigsaw" by one friend, a keen observer of others, charming and charismatic on stage while also elusive and private by nature. In respect to his orchestra, he is presented as an able manager who sometimes resorted to manipulating various members of a band he once described as "being made up of 18 maniacs." Ellington is also a driven and meticulous composer, clearly the most talented and prolific in all of jazz. One five-and-a-half-minute scene in this 18-minute chapter tells the story of his first appearance at Carnegie Hall, on January 23, 1943, at which he introduced his most ambitious work to date, a 44-minute tone poem called "Black, Brown, and Beige." This benefit concert was organized for Russian war victims who at the time were starving and destitute in the wake of the large-scale Nazi invasion. Ellington's intention was to move beyond the three-minute limitations of the popular song format to write an extended composition in three movements which "parallel[ed] the history of the Negro in America," according to narrator Keith David. The scene begins with a high contrast, black-and-white photograph of Ellington who appears elevated on stage in a dramatic silhouette, playing his piano. Ken Burns's distinctive stylistics are on full display, incorporating both those storytelling priorities of the classical Hollywood movies that he admired so much as an adolescent growing up in Ann Arbor with the more realistic and socially responsible aesthetics of the documentary tradition that he adopted years later as a college student and young professional. Reflecting these seemingly antithetical influences, Burns's account of Ellington's concert develops concurrently along narrative and thematic lines. Much of the story is relayed by a former trombone player in Ellington's orchestra, John Sanders, who provides a running commentary on the plot line of "Black, Brown, and Beige." In a parallel track, Burns and his creative team assemble a historically compelling montage of archival stills that aptly illustrate the three movements of the composition.

Part one, "Black," evokes slavery, as Ellington's musical references include both traditional work songs and the blues, over four vintage tableaux of plantation life from the cotton fields to the church. Part two, "Brown," is more joyous and up tempo in tone, alluding to those African Americans who fought for freedom during the Civil War, as five additional photographs complement this movement by showing blacks as federal soldiers staring purposefully into the camera, reflecting their own firm resolve in the pursuit of emancipation. Part three, "Beige," ushers the musical narrative into the twentieth century with an aggressively jazz-styled finale which captures the spirit and complexity of the modern era. "[Ellington] expressed all the movements of the Negro . . . northward," recounts Sanders, "Chicago, Philadelphia, Washington, and New York, and gets into the full stream of urban life and living."

Although the music takes precedence, as is customary throughout *Jazz,* the period imagery in "Beige" assists in both telling the story and visually documenting the cultural moment. This section specifically starts out with a slow pan to the right across nine black men, women, and children in farm clothing who are all standing by a river looking off into the distance. The next cut transforms this rural milieu into an energetic montage of eight archival photographs showing an assortment of active and self-assured African Americans now apparently at home in the city: two boys smiling and walking down a sidewalk; a dapper young man dressed in a stylish suit with a fedora cocked to one side; a couple in their car wearing fur coats; a boy posing in front of an automobile; two policemen in close-up; five children jumping rope on the street; two marines in full dress uniforms; and a little girl in a party outfit surrounded by other children and adults from the neighborhood. The entire scene culminates with a dramatic medium shot of Ellington exuberantly playing his piano, as singer Joya Sherrill affirms, "He was really a fighter for freedom. The limitations that had been put on blacks through the years was really unacceptable and so he was shouting musically—'we need to be free.'"

Ellington's celebrated debut of "Black, Brown, and Beige" before a Carnegie Hall audience that included the first lady of the United States, Eleanor Roosevelt, is just one of literally dozens of vignettes throughout *Jazz* where African American culture and the leitmotif of race relations share center stage with the music. Episode Seven, for example, is primarily concerned with events at home and overseas during World War II and, consequently, imbues Ellington's composition with an even greater power and poignancy when viewed within the context of one million black soldiers serving in a strictly segregated military in which even the "blood supplies were carefully separated by race." Only 20 minutes before the Ellington chapter, a brief but effective four-minute scene underscores the degree of racial prejudice and discrimination that still permeated the whole of the armed forces as well as the rest of American society. In a chapter entitled "Kill Jim Crow," violent outbreaks between black and white troops are presented as a routine fact of military life on bases across the United States. "Off-base black soldiers were harassed, beaten, barred from buses and even from restaurants where German prisoners of war were allowed to eat," narrates Keith David over an understated image of a black GI glancing upward at a "color waiting room" sign at a train station, "African Americans grew increasingly impatient with the hypocrisy of fighting bigotry abroad while tolerating it at home."

The arrival of bebop, too, assumes a far greater social as well as musical significance when observed alongside these other larger contextual developments in Episode Seven. In the summer of 1939, for instance, 19-year-old Charlie Parker is introduced in the opening scene, discovering "a new way to create a compelling solo based not on the melody of a tune, but on the chords underlying it." He and collaborator Dizzy Gillespie, who called Parker "the other part of my heartbeat," both explore this new style of playing, deepening their commitment to bebop during the early 1940s. A telling photograph of a young, hip, and well-dressed Gillespie appears early on in "Kill Jim Crow," as the camera tilts up slowly for 16 seconds, lingering first on the musician's expensive double-breasted suit, then his goatee and dark glasses, and finally his signature beret, before cutting back to a four-second close-up of his confident, smiling face. In this and similar shots in Episodes Seven and Eight, Parker's and Gillespie's "new assertiveness" are on full display, not only fueling their playing but also inspiring a nascent subculture made up mainly of bebop performers and their devotees. What the insiders of this new generational countercurrent had in common was a shared sense of rebelliousness and alienation, along with a growing attitude of racial pride for its African Americans members. "Black musicians started calling each other 'man,'" informs Keith David offscreen, "in part because they were so often called 'boy.'"

One of the most arresting scenes in all of *Jazz* arrives almost out of the blue as a kind of climax to Episode Seven. Pianist and composer Dave Brubeck is featured in a final chapter entitled "These Things Can't Happen," in which he recounts his World War II experiences in Europe during 1944 and 1945. For nearly five minutes Brubeck cheerfully recalls forming his integrated Wolfpack Band in a segregated army, surviving the Battle of the Bulge, and entering Germany with George Patton's 3rd Army, as assorted swing standards by Benny Goodman, Count Basie, and Glenn Miller revivify the stock footage and selected stills that outline the allied victory from the landing at D-Day through the eventual Nazi surrender in Berlin. "All over the world, jazz is accepted as the music of freedom," proclaims an enthusiastic Brubeck (even adding as if to further convince his interviewer, Ken Burns, "it's more important than baseball!").

The tone of the scene then shifts direction as the pianist remembers arriving back in Texas after the war, where his black band mates, still in uniform, were summarily refused service at a local restaurant. The indignity of this memory spurs an even more shocking boyhood recollection as Burns frames the musician's rough-hewn face in a tight, unsentimental 47-second long close-up accompanied by Brubeck's own sparse piano solo playing softly in the background: "You know the first black man that I saw, my dad [who was a rancher] took me to see [a friend] on the Sacramento River in California and he said to his friend, 'open your shirt for Dave.'" The musician pauses momentarily on-camera to compose himself while choking back his tears before continuing: "There was a brand on his chest and my dad said, 'These things can't happen. That's what I fought for.'" Dave Brubeck's startling and soulful retelling and re-experiencing of his more than 70-year-old remembrance encapsulates, in brief, the full range of emotions—anger, outrage, anguish, compassion, and shame—that are all a continuing part of America's enduring legacy of slavery and segregation.

From a musical point of view, Episode Seven is chock full of other equally understandable though much more stylized artistic responses to the conditions of institutional racism in the United States. The aforementioned "Black, Brown, and Beige" by Duke Ellington is one such example, especially in the way that it elegantly captures the outlook of a certain age group of African Americans, as is Charlie Parker's flight into bebop for another, younger generational perspective. Just as jazz became an international symbol of American democracy during World War II, the music also acted as a potent form of resistance, empowerment, and aspiration for those looking for a similar kind of freedom at home. All ten episodes of *Jazz*, in fact, emphasize the foundational importance of African American composers and musicians to the history of the music, as well as the centrality of blacks to the American experience in general. Ken Burns, in the end, is invariably liberal on most social issues, as is evident once again in his handling of race in this particular miniseries, while also being traditional in his broader conception of an America eventually coming together despite its many continuing conflicts based on color and cultural diversity.

Jazz overall is fully representative of Burns's work as a whole at midcareer. This ambitious multipart documentary fits easily into a long line of Ken Burns's films, all of which confirm certain aesthetic and ideological priorities, honed over a quarter-century of creating and producing 16 straight independent television specials without a false start. A final look at *Jazz* underscores five of the more salient characteristics that all his made-for-TV histories have in common. Specific examples focus on the way in which Burns and his creative team portray their chosen center of gravity, Louis Armstrong, throughout the mini-series.

Jazz, *first and foremost, reveals a seamless, solidly constructed plot structure which reflects Burns's partial integration of the Hollywood classical style into his methods as a popular historian.* Even at almost 19 hours (and more than 150 years of American history), this miniseries exhibits an epic storyline overflowing with historical people, places, and events. At the same time, its extended narrative design is virtually transparent to most of the viewers who are following along on TV, episode by episode, through the full extent of its chronological and biographical structure. Burns and his colleagues, in this way, strive for the seemingly realistic presentational strategies of Hollywood which involve telling stories in a straightforward if formulaic manner, relying heavily on one or more star protagonists—for example, Louis Armstrong and Duke Ellington—and employing the elements of film form—camerawork, editing, sound—as their means of always keeping the mechanics of the plot line hidden from millions of viewers.

"I am the audience," Burns readily admits, "if I have one gift, I think it is I have an ability in the editing room to be my audience's representative."[41] Nowhere is this trait more apparent than in the producer-director's ability to transform history into narrative and, thus, wake the dead by bringing select historical characters back to life within the confines of a well-told story about

America's past.[42] The mediated presence of Louis Armstrong, for instance, lights up the screen over and over again via photographs, film clips, and classic sound recordings of his voice and trumpet. He evolves from a street urchin to a jazz pioneer to a soloist extraordinaire to a globally recognized entertainer and good will ambassador for the United States government over the course of ten successive episodes. Armstrong's life in *Jazz*, accordingly, is couched within a complex mythic framework that characterizes him as a cross between a poor black Horatio Alger on one hand and music's coming messiah on the other. In this latter role, Louis Armstrong emerges from the humblest of beginnings, deep in the American South during the earliest years of the twentieth century, to become the one "chosen to bring the feeling and the message and the identity of jazz to everybody . . . all over the world," proclaims Wynton Marsalis in the opening moments of Episode Two. In true Hollywood fashion Marsalis adds, "he's the embodiment of jazz music," fusing *Jazz*'s transcendent hero with a classically structured narrative that is at once easy to follow, invisible to the eye, and fully consistent with Burns's distinctive stylistic approach as a historical filmmaker.

Second, the more specialized and scholarly subject matter of Jazz *is framed in lay terms for a general audience.* Ken Burns and most of his creative team began the *Jazz* film project as newcomers to the music themselves, taking years to gradually get up to speed on the topic's vast scope and broader significance. With this deeper appreciation "we didn't want to shortchange the depth, sophistication, and complexity of the music," acknowledges Lynn Novick, but "we also were always aware of the fact that we didn't expect our audience to have musical training—so we wanted to make sure that we instructed people about the music, and taught them how to listen, but never got too technical."[43] One particularly apt example that illustrates the spirit of this strategy is Burns's use of Matt Glaser in Episodes Two, Three, and Four to elaborate on Louis Armstrong's considerable talents. In the fourth installment especially, Glaser's remarks in the 18-minute 15-second chapter entitled "Mr. Armstrong" explain the inner workings of the music in both a richly informative as well as accessible fashion for the average viewer.

Glaser, a violinist who performed earlier on the soundtracks of *The Civil War* and *Baseball*, appears twice in "Mr. Armstrong." He begins the chapter speaking offscreen over 16 seconds of speeded-up stock footage of cars racing up and down city streets, people rushing along sidewalks, and a commuter train barreling around a concrete bend. Glaser describes the frenzied sounds of Armstrong's backup band in analogous terms: "It sounds like they are having a hard time coping with this fast tempo. The hectic tempo of the modern world is changing." Then Glaser's highly animated and expressive face fills the screen for an extended 60-second close-up as he provides a running commentary on the intricacies of Armstrong's trumpet solo in "Chinatown, My Chinatown": "free . . . completely relaxed . . . floating above [the rest of the band] . . . almost an aria." His is bravura performance, at once eccentric and instructive, offbeat and fully engaging: "Jazz has been dealing with this concept since Louis made this record, I mean, still to this day, now drummers and bass players and everyone can get into that groove. In those days he was the only guy to have this idea."

Burns brings Glaser back again nine minutes later to reprise his uniquely individualized contribution to the chapter—popping in momentarily like a quasi-familiar confidant—to offhandedly dissect the musical inventiveness of Armstrong's singing. He begins this second appearance in a 15-second voice-over as Armstrong is pictured still surrounded by nine other band members: "They are playing the melody in a very stiff, old-fashioned kind of way" (Glaser now onscreen hums, bobs his head, and gesticulates excitedly with his hands) "and then Louis comes in to show them a new way to [sing] the melody, articulated, completely free rhythmically, boiled down to one note, abstracted." (Cut to a high contrast, black-and-white photograph of a smiling Armstrong, eyes ablaze, singing into a microphone with his trumpet to the side in his right hand.) "Free, no time, all one note, he boiled down this complex melody to its essential impulse, then he decides to improvise" (as the camera gradually zooms in for 27 seconds on the image of Armstrong bringing the song and the scene to a joint conclusion).

All in all, Glaser's cameo illustrates the significance of Armstrong's playing and singing as no other strategy does throughout the entire miniseries. His face literally functions as a kind of musical road map for the viewer, orienting what to concentrate on and when. His intimate asides about the music as it is playing, moreover, clarify what specific aspects of the songs are important to *Jazz*'s early history and development and why. His observations are clear, insightful, never condescending, and always brimming with enthusiasm. Matt Glaser serves as genial guide to the music for all those nonspecialists watching at home. His appearance, more significantly, is indicative of the kinds of viewer identification strategies that Burns regularly incorporates into his work for the benefit of a widespread general audience.

Third, aesthetic and technical considerations influence the kinds of historical representations that are produced in Jazz. The design and construction of this "Mr. Armstrong" chapter is infused with an enthusiasm and passion for its subject that originates before all else with the music. Louis Armstrong's many gifts as a jazz innovator and entertainer are the main focal points of this section. Five of

Armstrong's songs actually punctuate the structure of "Mr. Armstrong" with each typifying in sequence another important aspect of this historical figure: "Chinatown, My Chinatown" (his virtuosity as a trumpet soloist); "Ain't Misbehavin'" (his crossover appeal to whites as well as blacks); "Dinah" (his electrifying and lovable stage presence); "Lazy River" (his remarkable skills and profound influence as a singer); and "Black and Blue" (his impact as a racial symbol and cultural icon). The amount of time accorded to each of these five tunes is generous, if not luxurious by motion picture standards, ranging from a high of three and a half minutes for "Chinatown, My Chinatown" to 90 seconds for "Dinah." "As a filmmaker," Burns explains, "it posed the most difficult challenge because, most of the time, music is background, the amplification of the emotion that's in a scene. Here we had to move that to the foreground and it had to work in many different ways."[44]

Before all else, it is crucial to remember, Burns is a film artist and a popular historian, not a scholar. He is, thus, applying his creative expertise as a producer-director to the historical material in *Jazz* in the hopes of reaching out to a broad-based national audience who—for the most part—is coming to the program with only a basic awareness of the topic. Burns, too, was in a similar position just six years earlier when he first committed to the production, armed only with an intense and wide-ranging curiosity to learn more about the music and its close relationship to the culture. "This could be looked at two ways," he maintains, "one the glass is half empty and you have some amateur coming in and doing a jazz history, but if the glass is half full, you have someone who has shared his heartfelt discovery of America's music with his countrymen."[45]

Fourth, Jazz *is yet another of Burns's documentary explorations into the nation's past that demonstrates his ongoing strategy as a popular historian to better define the present and to discover the future for himself, his immediate colleagues, and those audience members who regularly tune in to his television specials.* History for Ken Burns's films has primarily meant cultural biography, reflecting the producer-director's awareness and deep concern for contextual matters as well as the continuing influence that the social documentary tradition still exerts on him and his work. This stylistic feature of Burns's approach as a filmmaker manifests itself most clearly in the way he often addresses a whole host of relevant issues in his plot lines (e.g., race and ethnicity, gender and sexuality, urbanization, economic conditions), usually giving these agenda items their fullest expression through the life stories of his screen protagonists. Personal history, in this way, becomes the means by which larger national concerns are raised and examined, as Burns's documentary approach always establishes an intimate bond between the public interests of the country and the private experiences of his main characters.

Jazz likewise enacts this strategy by linking some of the more obvious trends in U.S. social history during the twentieth century to corresponding biographical details about Louis Armstrong. Burns and his creative team, for instance, cast a wide net in their original conception of the miniseries:

> In telling the story of jazz, our film will therefore shed light on a wide range of American cultural and historical eras and events that interact directly with the music, among them the harsh racial polarization of the 1890s and early twentieth century; the artistic and political ferment of the Harlem Renaissance; the exuberance of the Jazz Age; the liberalism of the New Deal; the anarchic, anti-authoritarian impulses of the Beat Generation; the emergence of a youth culture in the 1950s and 1960s; the hope, anger, and disappointments of the civil rights movement; and the search for identity and authenticity in the post-modern, self-referential 1980s.[46]

Scriptwriter Geoff Ward recognized in early 1995 that Louis Armstrong would be central to the evolving narrative development of *Jazz*. "And Ken came to see it very quickly too," he reveals, "as soon as he began to listen" to the music.[47] Burns corroborates Ward's point by acknowledging that "the surprise that was better than anything else was getting to know Louis Armstrong. At the beginning of the project, I knew him as a man with a big smile and a handkerchief, a transformer of popular songs like 'Hello Dolly' and 'What a Wonderful World.' But I didn't have any idea coming into this that he's the most important person in music in the twentieth century."[48] Armstrong, as a result, took on the leading role as the exemplar of jazz. He not only personifies the music, as might be expected, but he also came to embody the ascendancy of black culture and pride in the face of strict segregation during the jazz age all the way through to his public condemnation of President Eisenhower in the fall of 1957. This incident, which placed his career at risk, was coupled with his cancellation of a State Department tour of the Soviet Union because of the government's half-hearted attempts to desegregate Arkansas schools.

Armstrong's prominence as a symbol of imminent change in American race relations is first highlighted in his portrayal during the aforementioned "Mr. Armstrong" in Episode Four. Five minutes before the end of this chapter, following a spontaneous and unrestrained version of "Lazy River" where Louis ad-libs "boy am I riffing tonight—I hope," Keith David narrates over three glamorous publicity photos of the rising star: "In Harlem, young men took to carrying big white handkerchiefs because he flourished them on stage to mop his brow. Fans and fellow musicians alike began to copy his distinctive vocabulary. He was the first to refer to a musician's skills as his 'chops,' [and] the first to call people 'cats.'" As biographer Lawrence Bergreen affirms in *Louis Armstrong: An Extravagant Life:* "It was . . . apparent that he reveled in his blackness, in black language and bits of black experience that he tossed into his songs, and he wanted everyone to get hip to it along with him. He would sing, scat, kid the other musicians, and even address the instruments."[49]

"Mr. Armstrong" resumes with Louis's classic rendition of "Black and Blue," written originally as an ironic comedy number for the Broadway revue *Hot Chocolates.* Armstrong "transformed it," reports Keith David, "without a hint of self-pity, into a song about being black in a world run by whites." "Black and Blue," now recognized as one the first American protest songs dealing with racial discrimination (e.g., the lyric "my only sin, is in my skin"), is accompanied by a short but effective 95-second montage comprised of ten period snapshots and film clips documenting life under Jim Crow, including a young boy at a "colored" water fountain, a downtown laundry sporting a "We Wash For White People Only" sign, and a banner fixed high above a busy city street announcing, "A Man Was Lynched Yesterday." Armstrong's long-time bassist Arvell Shaw follows immediately, verifying onscreen that "Louis was the first man I heard say, you're black—be proud of it . . . he was saying that when it was very unpopular."

The chapter climaxes with a two-and-a-half-minute anecdote about a young University of Texas freshman, Charlie Black, who just happened to see Louis Armstrong perform at a local hotel in Austin one October evening in 1931. The experience would change his views about race forever, as his reaction is recounted by one of Burns's chorus of voices: "It is impossible to overstate the significance of a sixteen-year-old southern boy seeing genius for the first time in a black person . . . Louis opened my eyes wide and put to me a choice: 'Blacks,' the saying went, 'were all right in their place,' but what was the 'place' for such a man, and of the people from which he sprung?" Keith David then concludes this scene in a voice-over, as a class photograph of Charlie Black dissolves expectantly into a schoolroom of young black children perched at their desks, gazing directly, if a bit warily, into the camera: "Charlie Black went on to become Professor Charles L. Black Jr., a distinguished teacher of constitutional law at Yale. In 1954 he helped provide the answer to the question Louis Armstrong's music had first posed for him: he volunteered for the team of lawyers, black and white, who finally persuaded the U.S. Supreme Court in the case of *Brown v. Board of Education* that segregating school children on the basis of race and color was unconstitutional." Charles Black's closing testimonial strongly suggests that Armstrong's living example served as an inspiration for many, particularly a few select members of the next generation who would eventually take the fight for civil rights even further than had Armstrong and his contemporaries.

For Burns and company, Armstrong's genius goes well beyond the more limited spheres of music and entertainment; his larger significance, confirmed over and over again in *Jazz*, is the way he reinvented the aesthetic range, power, and popularity of jazz, thus prompting its widespread acceptance. This sensuous and suggestive music born in Storyville's brothels, shunned for years by polite society, had the unexpected side effect of bringing the races closer together, particularly in an era when total segregation was the norm throughout the United States. Even as late as the mid-1930s, black and white musicians were prohibited by social custom from playing together on stage; such intermingling was still a cause for concern and comment among many Americans well into the 1950s.

The sons and daughters of the World War II generation also harbored their own misconceptions. "I had a typical liberal sensibility that tends to suggest that those first white people who are successful [in jazz] ripped off the black community—the Bix Beiderbeckes, the Benny Goodmans, and the Paul Whitemans," admits Ken Burns in retrospect, "when in fact the opposite is true. These were people who risked everything because it was so socially unacceptable to find your teachers in the African American community. These are the vanguards of a country that's eventually going to change kicking and scream-ing."[50] In *Jazz*, the ultimate teacher of all is Louis Armstrong; and his greatest lesson is his willingness and ability to bridge categories: art and entertainment, musicians and fans, blacks and whites, and now another generation of admirers whose interest in him was given a vigorous and impassioned boost by the miniseries itself.

Viewing Louis Armstrong in such heroic and even transcendent terms is both a reformulation and a recovery of his image and legacy, especially from the frame of reference of a baby boomer filmmaker and his creative associates. More to the point, Wynton Marsalis listened mostly to "James Brown, Stevie Wonder, and Earth, Wind, and Fire" as a young teen. "Jazz was for old people," he noted in an outtake that was not used in the miniseries, "and if anybody heard of Louis Armstrong or Duke Ellington, 'they thought they were Uncle Toms.'"[51] Twenty-five years later, though, the slight hint of minstrelsy (i.e., his mugging, wide-eyed looks, and toothy smile) that is a part of Armstrong's performances of "Dinah" and other numbers in *Jazz*—which after all reflects the historical conditions into which he was born—is far overshadowed by his musical brilliance, his rhapsodic improvisations, and his personal ebullience on stage.

After his funeral in Episode Ten, Marsalis aptly summarizes onscreen: "Louis Armstrong's overwhelming message is one of love. When you hear his music, it's of joy. His music is so joyous. He was just not going to be defeated by the forces of life."

Jazz, like all of Ken Burns's historical documentaries, is part and parcel of a larger project that the producer-director began in 1977 and continued through all 16 of his film explorations into many of the more familiar chapters of American history. He is the filmmaker as bricoleur, searching through the flotsam and jetsam of the country's past, usually stored away in libraries and archives, retrieving and reordering this material into a historical made-for-television narrative for a new generation of citizen-viewers. Geoff Ward's recent contention that "[y]ou can't do American history and not do race," to cite an example, was an unthinkable proposition during Louis Armstrong's prime years as a musical innovator in the 1920s and early 1930s, and was later considered extreme by many in the United States as late as the 1950s.[52] Today, in stark contrast, it is a mainstream assumption, the stuff of popular history which Burns has made the "connecting thread in all of my films."[53]

"Every act of history is an act of interpretation," Ken Burns recognizes, "[s]o, to me, the greatest challenge is my responsibility to be honorable to the past, but also honorable to myself as an artist trying to determine why it is I'm drawn to this story of Louis Armstrong and not this person; and what is it in Louis Armstrong that I see indicative."[54] Armstrong personifies jazz for Burns and his creative team; he symbolizes a spirit of democracy that they are reconnecting with and reclaiming from an earlier generation, thus extending the musical and cultural legacy of their grandparents and parents into the present. The music, in this way, becomes another in an ever expanding series of subjects from American history that these particular baby boomer filmmakers have chosen to illustrate and reconfirm the enduring relevance and validity of the nation's majoritarian ideal, *e pluribus unum. Jazz* is Ken Burns's witness to the twentieth century, embracing his stylistic allegiances to both Hollywood storytelling and the social documentary tradition. As portrayed in this miniseries, jazz holds out the promise for a more enlightened, humane, and racially tolerant America in the future.

Fifth and, finally, Jazz *is a mixture of history and myth on film. It fully expresses the hybrid nature of popular history.* By made-for-television history standards, *Jazz* extensively maps out the who, what, where, when, and why of its topic over nearly 19 hours of running time. Although admittedly incomplete and framed largely within a neoclassical view of the genre, this miniseries stands as the most extensive history of the music on film so far. *Jazz*, most of all, represents a lucid articulation of Burns's vision as both an historian and a mythmaker, fine tuned during a quarter-century of producing and directing his own self-initiated projects. His stated goal is finding that "profound connection between remembering and freedom and human attachment. That's what history is to me."[55] Probing the past for Burns is his way of reassembling a filmic version of a more united future, often in the face of a deeply contentious and fragmented present.

Ken Burns's most influential mentors from the start, such as Jerome Liebling and David McCullough, were significantly artists and popular historians rather than academics (with whom he does regularly interact in an advisory capacity on all of his films). "I've had the good fortune to meet extraordinary human beings," Burns reminisces, "Lewis Mumford when I was working on a history of the Brooklyn Bridge. These ancient [Shaker] women in their 90s gave us and imparted so much of the beauty of their discipline and aesthetic life. Robert Penn Warren who helped me in my film on Huey Long. Shelby Foote with the Civil War series. Buck O'Neil who was a true mentor for the baseball series. And on and on and on."[56] Burns's relationship to history has always been far more intuitive and emotional than intellectual in orientation. "History is a way of self," he explains, "and I mean that at both the largest societal level, and at the deeply, intensely personal, psychological level."[57]

Put another way, Shelby Foote once told a reporter during the initial release of *The Civil War* that "'[a] fact is not the truth until you love it,' [poet John] Keats wrote in one of his letters, and that's the way it is with me, and with Ken, I believe."[58] Professional historians rarely talk this way; this emotive language is much more typical of the popular historian and the filmmaker. One scholarly critic, for instance, took particular issue with the obvious mythic reverberations throughout the miniseries, most evident in the celebratory remarks of journalists about their favorite jazz performers and in the poetic phrasings incorporated into the offscreen narration by Geoff Ward:

> *Jazz* would have been a substantially richer account of the music if Burns *had* paid more attention to the academics . . . . Instead of historically informed discussions of the music, its practitioners, and its reception, we hear, for example that Louis Armstrong's talent was "Godgiven," and that his playing could "make the angels weep." We are also told that Charlie Parker's genius is "unknowable," and that if Armstrong's solos were like poems, John Coltrane's were like the novels of Tolstoy. A little of this goes a long way, but it's the dominant rhetoric of the documentary.[59]

Ever since the beginning of his career, Burns has been a rapt enthusiast for the historical topics and figures he investigates, occasionally to a fault. In this miniseries, for instance, he refers to Louis Armstrong as "the most important person in music in the twentieth century, not just jazz. He is to music what Einstein is to physics, and what the Wright brothers are to travel."[60] On one level, such unrestrained hyperbole is evidence of "somebody [who] discovers something for himself, then you go a little overboard," according to noted jazz critic Nat Hentoff, who also serves as a commentator in the miniseries.[61] From another perspective, though, "mythical thought . . . works by analogies and comparisons even though its creations, like the 'bricoleur,' always really consist of new arrangements of elements."[62] Burns, in this context, is dead serious in proposing that Armstrong belongs in the same legendary company as the theoretical physicist and the inventors of sustained power-driven flight. In a much more fundamental way, he is actually championing a world view that seeks parity for the century's great achievements in art and culture alongside the longer accepted and more institutionalized benchmarks in science and technology. As a popular historian, he is, additionally, choosing people, places, and events from the country's past that are broadly representative of the priorities and values of his own peer group. Baby boomers, in this regard, are already deeply engaged in the process of finding their own generational position within the social and historical continuum that is the American experience. "I get kidded a lot," confesses Burns, "[t]hat keeps me in my place. I'm constantly reminded that people think this is a little overblown. At the same time, you can feel out there this palpable hunger for some sort of national self-definition."[63]

Burns utilizes the medium of choice for most Americans to convey his film vision to an audience of millions. Today television remains the center-piece of the nation's culture, even as it adapts to the ever expanding influence of the Internet. Practically all Americans from every walk of life watch lots of TV. For the entire two decades that Burns has relied on television as his professional home, the representative U.S. household has had its TV turned on for more than seven hours a day on average.[64] The characteristic baby boomer, in particular, will eventually spend nine full years in front of the television by the time he or she turns 65.[65] Burns thus works within the ideal medium to provide him with the widest possible access to his core viewership; at the same time, he regularly articulates his generation's deep ambivalence toward TV. "Television is one of the worst features of our daily life," he alleges, "it promotes disunion at a psychological, emotional, and personal level more insidiously than practically anything else in our culture. But it may also be, paradoxically, the instrumentality of our deliverance, a way in which we might find a common language and a common purpose."[66]

Different modes of communication are generally understood today as producing historic shifts in the way societies prefer certain forms of expression and knowledge over others. Oral communication flourished in nomadic cultures when people congregated around campfires in the evening after journeys to talk and swap gossip, anecdotes, and communal myths. Writing and later print communication slowly discouraged the widespread use of oral testimony and the heavy reliance on memory that it required. They, instead, "permit[ted] analysis, precision, and communication with future generations in a way not possible with the spoken word." "Human intimacy and community," however, have traditionally "best come through oral communication."[67] Modern forms of electronic communication, especially television, simulate some of the same functions once associated most with oral expression, such as fostering a common language, uniting people around similar stories, and providing an immediate and seemingly intimate forum for sharing collective memories. Such is the intent that Burns has always espoused for his TV histories—that is, sharing the American experience—in the hopes of promoting an increase in the country's cohesiveness at a time of unprecedented multicultural reassessment, redefinition, and change.[68]

Jazz, once again, fuses history with the epic storytelling power and reach of television to create a vision of America that is highly dynamic, creative, improvisational, and finally able to strike a workable balance between the needs of an increasingly diverse population and the nation's longstanding aspiration of forming one out of many. This majoritarian precept, *e pluribus unum*, has been called into particular question during the so-called culture wars of the last 20 years that probably reached their peak in the early to mid-1990s but, nevertheless, are still exerting a profound influence on the character and disposition of American life even today. This ongoing cultural dispute really has less to do with conflicts between formal political parties and religious sects than with the inevitable clash of competing and sometimes intertwined ideological world views.

The culture wars in this more fundamental sense are a sign of the dramatic realignment that is now taking place throughout American society, in which perspectives informed by race, ethnicity, gender, class, and generational position offer vastly different prescriptions for what the country's values should be and, consequently, what direction the nation should take in the future. Often the concept of the culture wars is oversimplified to be merely a stereotypical clash between proponents of two opposing systems of belief: those that assume that moral and ethical tenets are absolute and unchanging, and those that believe they are relative to the conditions and times in which we live. Most Americans, however, are much more eclectic in their various social, cultural, political, and religious convictions and affiliations and, therefore, far less dogmatic than this simple polarity suggests. Since 1980, at least, the culture wars have boiled over into a period of fierce partisanship and divisiveness throughout the country, based mainly on a struggle to define what America means and what it should be. Burns's historical documentary vision, more specifically, was shaped within and addresses this larger overriding context.

Ken Burns's television histories have always demonstrated his special aptitude for mediating traditional boundaries based on craft (film, photography, television), style (Hollywood and the social documentary), historical vantage points (the stories of the past with the concerns of the present), and separate generational perspectives (the cohort who came of age during World War II and their baby boomer children). His documentary vision also walks a thin line between competing ideological constituencies as he attempts to augment the nation's more traditional majoritarian view of itself by including and integrating chapters of the nation's history that have been marginalized and mostly forgotten, and recognizing them as essential to what the American experience really was, what it means today, and how it can be even more democratic and inclusive in the future.

Jazz clearly operates in this vein as yet another historical and mythic extension of Burns's vision. Myth, in this regard, is not meant in the old-fashioned sense of referring to a story that is false like a fairy tale. It is, instead, a fact-based, emotionally resonant facsimile of the actual event, produced under the usual limitations of living memory and the existing record. This miniseries, for example, is the result of a six-year commitment by the producer-director and his creative team to research and reassemble as responsible and valid a historical narrative as possible. David McCullough, one of Burns's earliest mentors, clarifies the essence of this particular process of historical mythmaking with his favorite quotation from the nineteenth century French painter Eugene Delacroix, "What I require is accuracy for the sake of imagination." McCullough explains that "[t]he accuracy has to be in the research. But the writing of history and biography is very largely an imaginative act. It doesn't mean by imagination that you are making things up. It means that you are putting two and two together and you are putting yourself into the scene and the time as an act of empathy and imagination."[69]

Ken Burns adopts this kind of intense personal relationship with all the major historical figures he chooses to bring to life on film. Burns established an especially close attachment to Louis Armstrong who emerged as the structural and moral apotheosis of the entire miniseries. As the main protagonist, Armstrong is never twisted out of shape to symbolize something that has no historical basis in fact. His image, however, is enlarged to signify much more than it ever has in the past. As expected, the Louis Armstrong of *Jazz* looms above all the rest as "the first great protean genius of the music." Even more significantly, though, his presence also transforms into the veritable second coming of "jazz's *redemptive future promise.*"[70] Armstrong's generosity of spirit coupled with his innate warmth, friendliness, and sense of courtesy and respect in his dealings with others comes to personify the life-affirming influence of the music. Louis Armstrong thus functions as a kind of mythic reminder that the nation's racial difficulties can be mediated and common ground eventually achieved.

In the end, Ken Burns's focus on history and the extended lengths of his prime-time specials—telecast without commercial interruptions—all implicitly challenge the idea of television as entertainment that most Americans take for granted. He further underscored the therapeutic potential of his historical and mythic narratives when he explained in an interview that

> when an individual has an identity crisis, you send him to a therapist. The first
> thing a therapist wants to know is, where did you come from, who are your
> parents, that is to say, what is your history? In the exploration of the nooks and
> crannies of that [personal] history, however painful, are the seeds of healing;
> and history, if done in that open-ended way, can be about something tragic
> and divisive like race relationships, apartheid, segregation, lynchings, and at
> the same time, through an honest and, I would say, artistically presented
> investigation, it has the possibility of healing by shedding light on the traumas
> that exist just not in our individual psyches but in our national psyche.[71]

All in all, history for Burns is deeply personal in practice but public in its scope and significance. It represents a lifelong commitment by him as a filmmaker and as a popular historian. "The 'bricoleur' may not ever complete his purpose," concludes Lévi-Strauss, "but he always puts something of himself into it."[72]

THE PEOPLE'S HISTORIAN TAKES FIVE

> I never expected to be in this place. I started out as just wanting to be a
> filmmaker. But I've grown up, and I've learned how to doggedly pursue
> support. Your work doesn't mean much if it doesn't get seen.
>
> —*Ken Burns, 2000*[73]

The improbable nature of Ken Burns's career as a television producer and a film director took yet another unprecedented turn in the summer of 1999 when he signed an exclusive ten-year deal with General Motors: the automaker agreed to underwrite 35 percent of all his production budgets and 100 percent of any associated educational outreach efforts over the full duration of the contract. When this windfall arrangement was reported in the *New York Times* on June 16 of that year, the article began with the eye-catching phrase, "Contortions of envy may well ripple through the documentary filmmakers' world tomorrow."[74] This lead obviously emphasizes Burns's unique position as the first among equals within the professional ranks of the always struggling and chronically under-funded independent production sector. Ever since *The Civil War,* noted a reporter for *Life* during the initial release of *Baseball* in September 1994, "Burns has been a man with one foot in lights-out Walpole, and one in the flashbulb world of American celebrity."[75] His name recognition in TV households across the country remains as high today as it was when he originally became documentary filmmaking's first breakout producer-director in the fall of 1990.

No maker of prime-time television histories has ever been as popular with a nationwide audience as Ken Burns. Few have also been as celebrated. He has received nearly every possible award that is relevant for him to win; his only contemporary rival in this regard is David Grubin. Burns had a day designated in his honor by the governor of New Hampshire in October 1990, while that state's Senator Warren Rudman read praise for the quality of his work into the Congressional Record. Since the early 1990s, Burns has literally made dozens of appearances on television, radio, and even in films (such as his cameo as Hancock's aide in *Gettysburg* [1993]), as well as seriously entertaining an offer to coproduce a dramatic feature on Jackie Robinson with the Merchant-Ivory team for Disney's Hollywood Pictures in late 1994 and early 1995. He occasionally receives letters addressed simply "Ken Burns, New Hampshire" and has now lent his image to the advertising of products outside the purview of his own projects, such as New Hampshire's Stonyfield Farm yogurt, although characteristically, his compensation for this service was a donation made in his name to the Louis Armstrong House Educational Foundation.

Burns also has been a public spokesperson for a wide variety of issues and causes throughout the last decade, such as opposition to the Disney Corpora-tion's plans to build a historic theme park beside the Manassas Battlefield in Virginia and his periodic appearances on Capitol Hill to speak in front of various subcommittees on behalf of the Corporation for Public Broadcasting, the Public Broadcasting Service, the National Endowment for the Humanities, and the

National Endowment for the Arts. Burns has, moreover, been a regular visitor to the White House throughout the 1990s, beginning with an invitation from President George H. Bush after the debut of *The Civil War,* and including special screenings of *Thomas Jefferson* in early 1997 and *Lewis & Clark: The Journey of the Corps of Discovery* on November 14, 1997 to kick off the White House Millennium Program—to "help us honor our past and imagine the future," said President William J. Clinton at that evening's ceremony, "and for that I turn to the incomparable Ken Burns."[76]

Burns is arguably the most recognizable and influential historian of his generation. His significance stems from his being a historical storyteller who reaches his largest audiences through television. He has seized the attention of large segments of the country's viewing public by the subjects he chooses and the way he presents them. Burns, overall, articulates a version of the country's past that conveys many widespread assumptions about the historic traditions and the mythic character of the United States. "I am really an evangelist," he argues, "not just for history but for a kind of a vision of who we might be."[77] In discovering his own personal niche between the academy and the general public, Ken Burns has realized his life's work as the historian of choice for the many millions of Americans who regularly tune in to his stories on television. "I've sort of been lucky that way in my professional life," he admits, "It's funny [how] I've hit the zeitgeist accidentally" over the years.[78]

Jazz was further evidence of this point. The series debuted to an estimated 13 million viewers on January 8. Nielsen averages put *Jazz* at a 3.6 household rating and a 6 percent share of the national audience for the run of the ten episodes during four successive weeks in January 2001.[79] These percentages are double the customary public television averages, translating into approximately 23 million viewers when calculated over the entire length of the miniseries. Initial sales figures for *Jazz's* ancillary products were similarly auspicious. Advance orders for the companion book from Knopf exceeded 200,000, while the five-CD, 94-track boxed set from Columbia/Verve, *Ken Burns Jazz: The Story of America's Music,* was certified gold with more than 500,000 copies sold by late January.[80] Better than expected numbers were also tallied for the 22 single-disc anthologies of artists featured in the miniseries.[81] On the January 27 *Billboard* magazine Top Jazz Albums chart, 16 of the top 25 spots bore the Ken Burns imprint, and this trend continued through the rest of the spring. "Jazz sales were up 20 percent . . . compared to a 6 percent downturn in overall CD sales," reported Michael Kaufman, senior vice president of sales and catalog for the Verve Music Group, "[w]e're even seeing sales in Middle America."[82] "The Ken

Burns thing is incredible," added Verve president Ron Goldstein "[f]ortunately for us, Ken Burns was an established brand name . . . [t]he products were first-rate, and people believed in him."[83]

Burns's widespread popularity and commercial clout were ultimately the deciding factors that encouraged such long-time and intractable jazz competitors as Columbia/Legacy and Verve to share their catalogs in producing these new CD compilations. They were joined later by such labels as Blue Note, RCA, and Rhino, who contributed other hard-to-get recordings.[84] As had been the case six years earlier with *Baseball*, Florentine Films/American Documentaries and General Motors, with the addition of Dan Klores Associates, designed and enacted a marketing campaign that became the new professional standard of excellence in promoting a public television program. "Working with a team of some 100 communications and merchandising specialists," reported business weekly *Crain's*, "Mr. Burns, his corporate sponsor General Motors Corp. and a public relations firm have assembled the kinds of partnerships usually made for major Hollywood productions."[85]

The *Jazz* team combined forces with retailers such as Borders, Barnes & Noble, and Tower Records, who all utilized special displays for the companion book and the two dozen supplementary CDs; 3,000 Starbucks coffeehouses nationwide were enlisted to promote the miniseries and the accompanying CDs with music, banners, and viewer guides; the National Basketball Association, starting in December, hosted publicity events during game half times that included short video previews of the series and live jazz performances.[86] Two state-of-the-art websites were constructed around the miniseries at PBS and the musical compilations at Verve. Furthermore, the main charitable recipient of the resulting revenues for this corporate coalition was the United Negro College Fund, and middle school teachers received more than 75,000 educational viewers guides as well as specially edited videos and CD samplers for use in public and private schools across the country.[87]

Ken Burns once again became the celebrity centerpiece of the whole promotional juggernaut that was built around *Jazz*. He actually began to make occasional appearances specifically for his new miniseries as early as April 2000 and continued these intermittent previews of the program throughout much of the summer. The primary phase of the marketing campaign began in earnest on Wednesday, September 6, however, with a major press conference at the celebrated New York jazz club Birdland.[88] From that point onward, Burns's first responsibility was no longer the editing of another upcoming production, *Mark Twain*, but being the featured spokesperson for *Jazz*, crisscrossing the country several times, publicizing his latest miniseries as comprehensively as possible until its debut four months later on January 8, 2001. "[Ken Burns is] an excellent salesman," observes Phil Risinger, manager of brand activities at General Motors, "He's very eloquent, in a way that captures everyone's attention. And he grows, in a business sense, with each project."[89]

Burns's celebrity and his high-profile commercial activities are the source of some skepticism among his professional peers (mostly off the record), and undoubtedly affect the tenor of sporadic reviews from both the scholarly and popular press (although to what degree is impossible to say). His own reaction to his heightened visibility and his many business obligations is to admit that "there are many dangers to this work and success or celebrity or fame . . . is a big problem, but it is not that big compared to being true to the work."[90] He also is more philosophical than ever about the criticisms he invariably receives as each new documentary special debuts on television: "I've been quite fortunate the first seven or eight films that I made were sort of universally praised and somewhere toward the end of that I developed a thick skin anyway, and so by the time people would level criticisms at the films I was prepared for it, and there is always a germ of truth even in the most vitriolic criticism that you get, and so it is important to look at it and to see it for what it is. I think it strengthens the work along the way."[91]

Subsequent critiques of *Jazz* in particular centered chiefly around three principal issues: the truncated nature of the narrative, the neoclassical view of the genre, and Ken Burns's emotive approach to historical storytelling. The first two matters are largely interdependent and related to what Ben Ratliff of the *New York Times* means when he calls the miniseries "a middle-of-the-road version of jazz history . . . in its selection and depiction of individual figures."[92] As noted earlier in this chapter, Ken Burns and his creative team focus most of their attention on the 43 years (1917 to 1960) comprising Episodes Two through Nine; this decision reflects both the great-man imperatives of the narrative and the liberal pluralist leanings of the filmmakers. This historical rather than future oriented view of jazz also coincides with the neoclassical school of the genre most associated with Wynton Marsalis, the composer, musician, and artistic director of the Jazz at Lincoln Center series, and his two closest mentors in augmenting his background knowledge about the history of the music, Albert Murray and Stanley Crouch. "For many people, [Marsalis] be[came] the symbol of the rebirth of mainstream jazz," championing the founding fathers of the music in the face of stiff opposition from other outspoken factions in the jazz community who each advocated their own contemporary alternatives drawn literally from "hundreds of different creators out there, all pursuing their own paths, a number of which may turn out to have lasting merit."[93]

Marsalis, as a result, embraced the role of lightning rod, "insist[ing] that neither most fusion nor much of the avant-garde should ever have been considered jazz, since they didn't swing and lacked blues feeling."[94] He also became a highly controversial figure in the often contentious and insular world of jazz by "denounc[ing] the cult of the 'new,' scoffing at the notion that change necessarily means progress, that one kind of jazz is ever an 'advance' over another."[95] Unwittingly or not, Burns stepped into this debate when he selected Marsalis as his senior creative consultant. Even before the editing phase of the miniseries began, salvos from jazz aficionados were already being launched at the program as just "more propaganda for the revisionist history of jazz according to Wynton Marsalis."[96] For Burns's part, his historical inclinations dovetailed perfectly with Marsalis's viewpoint from the beginning, although they too had a few disagreements mostly over Burns's decision to merely describe rather than strongly criticize fusion and the avant-garde in Episode Ten.

After a while, both Geoff Ward and Burns grew accustomed to the controversy. Ward typically dealt with it wryly, referring to the "jazz community" as a "dysfunctional family," and promising that "[i]t's not a documentary that will make jazz experts particularly happy. There's something missing for everyone."[97] Burns was predictably more direct in his assessment that "we made this film for a broad national audience. We didn't make it to please jazz critics. We made it to please a broad national audience because this is our birthright. This is who we are."[98] "[Burns i]s a great filmmaker in the documentary style, and his films attest to that," observes Marsalis in summarizing his working relationship with the producer-director, "He's a tireless worker. He's in the top percentile of serious people—that midnight-oil-burnin' seriousness, where you can't go to sleep unless it's right. The thing I really respect about him is that he'll listen to what you have to say, but he makes his film the way he wants it to be."[99]

Ken Burns's most fundamental aim as a popular historian, overall, is to explore the experience of history rather than to provide a detailed description and analysis of his subject, much to the chagrin of his detractors in the academy. "You may be moved by Ken Burns's images in *Jazz*. I certainly was," wrote one jazz scholar for example, "[u]ltimately, however, the program eschews nuance and complexity for the emotive gesture, even if it sometimes means saying nothing about issues that beg to be lucidly discussed."[100] Professional history tends to privilege reasoned discourse over artistic ways of seeing the world. Even as historical films and television programs are becoming more common in academe, they are still mostly relegated to supporting roles behind the supposedly more substantial findings that are first established by delineating and then evaluating the empirical data. Film documentaries, in this way, are perceived to be soft on facts, figures, and analysis; TV programs, specifically, are considered the most ephemeral and untrustworthy of all historical forms.

In retrospect, Ken Burns's whole career has been an intensely personal quest into the subjective nature of the past. The historical questions that tend to interest him the most, as a so-called emotional archaeologist, are the ones that ask what it really felt like to be part of certain events in the past; what was the experience of inhabiting such long ago places and being from another time; what did people back then fear the most; what did they desire; and what were their dreams similar to what Americans think about and hope for today.[101] Burns, the popular historian, has spent a quarter-century of nonstop professional effort creating and refining film techniques that address these kinds of questions— however imperfectly. Burns, in the process, reverses the academic hierarchy, trusting first the lessons found in art (photographs, film clips, period music, paintings, etc.), before turning to the scholarly record to fill in the details of his vision of American history. His is admittedly a speculative approach; but then again, filmmakers and professional historians alike are all amateurs when it comes to detecting the human traces of lives once lived among the emotional resonances of the past.

Ken Burns finally took a break from producing and directing documentaries through much of 2001 to just relax, read, and spend more time with family and friends. This well-earned sabbatical was his first extended leave from historical filmmaking since devoting his full-time attention to making *Brooklyn Bridge* back in January 1977. Even during a short vacation that he took away from his many pending projects for public television in 1995, Burns, along with old friends and long-time collaborators Roger Sherman, Buddy Squires, and Wendy Conquest, produced a 31-minute filmed meditation on the close interrelation-ship between "the innate call in each human being to search" and the creative process. The film featured painter William Segal at the ancient French basilica *Vézelay* (which is also the title of the work). Burns returned home afterward to Walpole, New Hampshire, to resume his own personal exploration of America's past in just the kind of town that suggests an earlier time. "I think living in New Hampshire," he confides, "free from the . . . marketplace and competition of the big filmmaking cities, reminds me what I'm here to do—which is work. That is what I do as much as anybody I know."[102]

Before taking a year off, though, Burns did make preliminary plans for what he would be doing upon his return in early 2002. "Having a long-term commitment" from General Motors, he explains, "allows you to do some real long-term planning." Burns began by outlining five more projects, thus extending the eventual scope of his *American Lives* series from five to ten episodes. Future subjects include *Horatio's Drive* (about Horatio Nelson Jackson who crossed the country by car on a bet nearly 100 years ago), *World War II* (which features several prominent historical figures), boxer Jack Johnson, and probably Dr. Martin Luther King, Jr. and Elvis Presley. He has also completed a film treatment for an ambitious miniseries on the history of the national parks system which he intends to coproduce with his close friend, colleague, and fellow Walpole resident Dayton Duncan. Expectations are sure to be high when Burns's future historical documentaries eventually premiere on television screens all over America. No longer will he be able to sneak up on the country's viewers as he once did with *The Civil War* back in September 1990.

Lightning rarely strikes twice in any filmmaker's career. Burns's personal stamp, nevertheless, is evident throughout the closely interconnected worlds of prime-time television and historical film and video production. To begin with, the originality of Burns's well recognizable and distinctive style has now become normative for an entire generation of historical documentarists, intimating by the ubiquity of examples that imitation is still the sincerest form of flattery. Second, the resurgence in the TV documentary that he helped initiate more than a decade ago continues to flourish, marked by a new wave of important young directorial voices and innovative nonfictional works, especially when compared with the far more predictable and heavily formatted releases emanating from the independent fiction filmmaking sector over the same time period. Third, and most improbably when taking into account the virtual absence of historical programming on the small screen throughout most of the 1960s and 1970s, television is today the principal means by which most Americans learn about history. Just as TV has profoundly affected every other aspect of contemporary life—from the family to education, government, business, and religion—the medium's historical portrayals have similarly transformed the way tens of millions of viewers think about the past. In this regard, no one has been more prolific as a producer-director, set a higher professional standard, and had a greater impact on rekindling America's ever increasing interest in all things historical than Ken Burns.

Notes

CHAPTER 1

1. Ken Burns, "The Documentary Film: Its Role in the Study of History," text of speech delivered as a Lowell Lecture at Harvard College, 2 May 1991, 6.
2. Ken Burns, telephone interview with the author, 18 February 1993.
3. Ken Burns, interview with the author, 27 February 1996.
4. Neal Gabler, "History's Prime Time," *TV Guide*, 23 August 1997, 18.
5. Shelby Foote, *Civil War: A Narrative (Fort Sumter to Perryville, Fredericksburg to Meridan, Red River to Appomattox)*, 3 vols. (New York: Random House, 1958-1974); David McCullough, *The Great Bridge: The Epic Story of the Building of the Brooklyn Bridge* (New York: Touchstone, 1972); and Michael Shaara, *The Killer Angels* (New York: Ballantine, 1974).
6. Ken Burns, "Four O'Clock in the Morning Courage," in *Ken Burns's The Civil War: Historians Respond*, ed. Robert B. Toplin (New York: Oxford, 1996), 157.
7. Ken Burns quoted in "A Filmmaking Career" on the *Thomas Jefferson* (1997) website at <http://www.pbs.org/jefferson/making/KB_03.htm>.
8. The $3.2 million budget for *The Civil War* was comprised of contributions by the National Endowment for the Humanities ($1.3 million), General Motors ($1 million), the Corporation for Public Broadcasting and WETA-TV ($350,000), the Arthur Vining Davis Foundation ($350,000), and the MacArthur Foundation ($200,000). General Motors also provided an additional $600,000 for educational materials and promotional outreach.
9. Matt Roush, "Epic TV Film Tells Tragedy of a Nation," *USA Today*, 21 September 1990, 1.
10. See Lewis Lord, "'The Civil War': Did Anyone Dislike It?" *U.S. News & World Report*, 8 October 1990, 18; *Civil War Illustrated*, July/August 1991; *Confederate Veteran*, January/February 1991, March/April 1991, July/August 1991; and Toplin, *Civil War: Historians Respond*.
11. The consultants as listed in the credits are Shelby Foote, Barbara J. Fields, C. Vann Woodward, Don Fehrenbacher, Stephen Sears, William McFeely, James McPherson, Bernard Weisberger, Mike Musick, Richard Snow, Eric Foner, Stephen B. Oates, Robert Johannsen, Tom Lewis, William E. Leuchtenburg, Daniel Aaron, Charles Fuller, Charley McDowell, Ira Berlin, Gene Smith, Robert Penn Warren, Jerome Liebling, Dayton Duncan, and Amy Stechler Burns.

12. Some of the more prominent critiques of *The Civil War* focus on errors in detail, the way the series abridges the origins of the war and the later matter of reconstruction, and the condensation of other complex issues, such as policy making and the formation of public opinion. For additional disagreements in interpretation see Jerry Adler, "Revisiting the Civil War," *Newsweek,* 8 October 1990, 62; Jane Turner Censer, "Videobites: Ken Burns's 'The Civil War' in the Classroom," *American Quarterly* 44.2 (1992): 244-54; Mary A. DeCredico, "Image and Reality: Ken Burns and the Urban Confederacy," *Journal of Urban History* 23.4 (1997): 387-405; Ellen Carol DuBois, "The Civil War," *American Historical Review* 96.4 (1991): 1140-1142; A. Cash Koeniger, "Ken Burns's 'The Civil War': Triumph or Travesty?" *The Journal of Military History* 55 (April 1991): 225-33; David Marc and Robert J. Thompson, *Prime Time, Prime Movers: From I Love Lucy to L.A. Law—America's Greatest TV Shows and the People Who Created Them* (Boston: Little, Brown, 1992), 307; Robert E. May, "The Limitations of Classroom Media: Ken Burns' Civil War Series as a Test Case," *Journal of American Culture* 19.3 (1996): 39-49; Hugh Purcell, "America's Civil Wars," *History Today* 41 (May 1991), 7-9; and Mark Wahlgren Summers, "The Civil War," *Journal of American History* 77.3, 1106-07.

13. Koeniger, "Triumph or Travesty?" 233. Like many film scholars before him, Louis Giannetti writes, "*Birth* is a diseased masterpiece, steeped in racial bigotry," in *Masters of the American Cinema* (Englewood Cliffs, N.J.: Prentice-Hall, 1981), 67. This critical ambivalence about *Birth of a Nation* in general film histories dates back to Terry Ramsaye, *A Million and One Nights* (New York: Simon & Schuster, 1926), Benjamin Hampton, *A History of the Movies* (New York: Covici, Friede, 1931), and Lewis Jacobs, *The Rise of the American Film* (New York: Harcourt, Brace, 1939).

14. John Milius, "Reliving the War Between Brothers," *New York Times,* 16 September 1990, sec. 2, 1, 43.

15. Russell Merritt, "Dixon, Griffith, and the Southern Legend: A Cultural Analysis of *Birth of a Nation,*" in *Cinema Examined: Selections from Cinema Journal,* ed. by Richard Dyer MacCann and Jack C. Ellis (New York: Dutton, 1982), 167, 175.

16. Merritt, "Dixon, Griffith, and the Southern Legend," 166.

17. Statistical Research Incorporated (Westfield, N. J.), "1990 Public Television National Image Survey," commissioned by PBS Station Independence Program, 28 September 1990, 2.1-2.8.

18. "CBS, PBS Factors in Surprising Prime Time Start," *Broadcasting,* 1 October 1990, 28; "Learning Lessons from 'The Civil War,'" *Broadcasting,* 8 October 1990, 52-53; Richard Gold, "'Civil War' Boost to Docu Battle," *Variety,* 1 October 1990, 36; Bill Carter, "'Civil War' Sets an Audience Record for PBS," *New York Times,* 25 September 1990, C17; Jeremy Gerard, "'Civil War' Seems to Have Set a Record," *New York Times,* 29 September 1990, 46; and Susan Bickelhaupt, "'Civil War' weighs in with heavy hitters," *Boston Globe,* 25 September 1990, 61, 64.

19. George F. Will, "A Masterpiece on the Civil War," *Washington Post,* 20 September 1990, A23.

20. David S. Broder, "PBS Series Provides a Timely Reminder of War's Horrors," *Sunday Republican* (Springfield, Mass.), 30 September 1990, B2; Haynes Johnson, "An Eloquent History Lesson," *Washington Post,* 28 September 1990, A2.

21. Harry F. Waters, "An American Mosaic," *Newsweek*, 17 September 1990, 68; Richard Zoglin, "The Terrible Remedy," *Time*, 24 September 1990, 73; Lewis Lord, "The Civil War, Unvarnished," *U.S. News & World Report*, 24 September 1990, 74.

22. David Thomson, "History Composed with Film," *Film Comment* 26.5 (September/October 1990), 12.

23. Tom Shales, "The Civil War Drama—TV Preview: The Heroic Retelling of a Nation's Agony," *Washington Post*, 23 September 1990, G5.

24. Monica Collins, "A Victory for 'Civil War,'" *Boston Herald*, 21 September 1990, 43.

25. "Ken Burns Wins First Lincoln Prize," *American History Illustrated* 26.3 (July-August 1991), 14.

26. Ken Burns received honorary degrees from the following eight institutions in 1991: LHD (hon.), Bowdoin College; LittD (hon.), Amherst College; LHD (hon.), University of New Hampshire; DFA, Franklin Pierce College; LittD (hon.), Notre Dame College (Manchester, N.H.); LittD (hon.), College of St. Joseph (Rutland, V.T.); LHD (hon.), Springfield College (Ill.); and LHD (hon.), Pace University.

27. Burns, interview, 18 February 1993.

28. "Ken Burns," *People*, 31 December 1990-7 January 1991, 46-47.

29. Daisy Mayles, "1990 Facts & Figures: Fiction Sales Outpace Nonfiction," *Publishers Weekly*, 8 March 1991, 20.

30. "Longest-Running Hardcover Bestsellers for 1991," *Publishers Weekly*, 1 January 1992, 34.

31. Mayles, "1990 Facts & Figures: Fiction Sales Outpace Nonfiction," 20.

32. Ken Burns, interviewed by David Thelen, "The Movie Maker as Historian: Conversations with Ken Burns," *Journal of American History* 81.3 (1994), 1050.

33. Eileen Fitzpatrick, "Burns Sues Pacific Arts for Back Royalties on Videos," *Billboard*, 22 October 1994, 9.

34. Burns, interview, 18 February 1993.

35. James M. McPherson, *Battle Cry of Freedom* (New York: Oxford, 1988), 865.

36. Edwin McDowell, "Bookstores Heed Call on Civil War," *New York Times*, 1 October 1990, D10.

37. Foote, *Civil War: Red River to Appomattox*, 1064; Lynne V. Cheney, "A Conversation with Civil War Historian Shelby Foote," *Humanities* 11: 2, March/April 1990, 8.

38. Burns, interview, 18 February 1993.

39. "*The Civil War:* Ken Burns Charts a Nation's Birth," *American Film* 15 (September 1990), 58.

40. Ken Burns, "In Search of the Painful, Essential Images of War," *New York Times*, 27 January 1991, sec. 2, 1.

41. Horace Newcomb, *TV: The Most Popular Art* (New York: Anchor, 1974); John Fiske and John Hartley, *Reading Television* (New York: Methuen, 1978); Richard P. Adler, ed., *Understanding Television: Essays on Television as a Social and Cultural Force* (New York: Praeger, 1981); Robert C. Allen, ed., *Channels of Discourse: Television and Contemporary Criticism* (Chapel Hill: University of North Carolina Press, 1987); David Bianculli, *Teleliteracy: Taking Television Seriously* (New York: Continuum, 1992); Horace Newcomb, ed., *Television: The Critical View*, 5th ed. (New York: Oxford, 1994).

42. Ken Burns, "Mystic Chords of Memory," text of speech delivered at University of Vermont, 12 September 1991, 14.

43. "*The Civil War:* A Television Series," grant application with appendices submitted to National Endowment for the Humanities, 14 March 1986, 9.

44. Burns in Thelen, "Movie Maker as Historian," 1035.

45. Daniel J. Boorstin, "The Luxury of Retrospect," *Life, Special Issue: The 80s* (Fall 1989), 37.

46. Burns in Thelen, "Movie Maker as Historian," 1037.

47. Burns, interview, 18 February 1993.

48. Burns in Thelen, "Movie Maker as Historian," 1037.

49. Burns, interview, 18 February 1993.

50. Burns in Thelen, "Movie Maker as Historian," 1040.

51. Bernard A. Weisberger, "The Great Arrogance of the Present Is to Forget the Intelligence of the Past," *American Heritage* 41.6 (September/October 1990), 99.

52. Ron Powers, "Glory, Glory," *GQ,* September 1990, 218.

53. The final draft of the script, dated 17 July 1989, is in the Ken Burns Collection at the Folklore Archives of the Wilson Library, University of North Carolina at Chapel Hill. *The Civil War* materials include all drafts of the script, all the filmed interviews with various scholars and experts (including outtakes), other footage, notes on decision making, test narrations, some financial records, correspondence, other related data.

54. Robert Sullivan, "The Burns Method: History or Myth-Making?" *Life,* September 1994, 42.

55. Cathryn Donohoe, "Echoes of a Union Major's Farewell," *Insight,* 5 November 1990, 54-55; and Susan Bickelhaupt, "Civil War Elegy Captivates TV Viewers," *Boston Globe,* 29 September 1990, 1, 5.

56. Burns, interview, 18 February 1993.

57. Ken Burns, interviewed by Thomas Cripps, "Historical Truth: An Interview with Ken Burns," *American Historical Review* 100.3 (1995), 752.

58. Burns, "Mystic Chords of Memory," 11.

59. Burns, interview, 18 February 1993.

60. Censer, "Videobites," 245.

61. DuBois, "The Civil War," 1140-41.

62. Dylan Jones and Dennis Kelly, "Schools Use Series to Bring History to Life," *USA Today,* 1 October 1990, 4D; Burns, "Mystic Chords of Memory," 16.

63. Peter Novick, *That Noble Dream: The "Objectivity Question" and the American Historical Profession* (Cambridge: Cambridge University Press, 1988).

64. Lawrence W. Levine, *The Unpredictable Past: Explorations in American Cultural History* (New York: Oxford, 1993), 8.

65. Burns, interview, 18 February 1993.

66. "*The Civil War:* Ken Burns Charts a Nation's Birth," 58.

67. Burns, interview, 18 February 1993.

68. Ken Burns, "Thoughts on Telling History," *American History Illustrated* 26.1 (1991), 27.

69. Adler, "Revisiting the Civil War," 61.

70. The literature encompassing memory studies is vast and found in many disciplines. A few well-known sources to start with are: Paul Fussell, *The Great War and Modern Memory* (New York: Oxford, 1989); Michael Kammen, *Mystic Chords of Memory: The Transformation of Tradition in American Culture* (New York: Vintage, 1993); Jacques Le Goff, *History and Memory,* European Perspectives (New York: Columbia University

Press, 1996); Bernard Lewis, *History: Remembered, Recovered, Invented* (Princeton, N.J.: Princeton University Press, 1975; George Lipsitz, *Time Passages: Collective Memory and American Popular Culture* (Minneapolis: University of Minnesota Press, 1990); Michael Schudson, *Watergate in American Memory: How We Remember, Forget and Reconstruct the Past* (New York: Basic Books, 1992); and Barbie Zelizer, *Covering the Body: The Kennedy Assassination, the Media, and the Shaping of Collective Memory* (Chicago: University of Chicago Press, 1992).

71. David Thelen, "Memory and American History," *Journal of American History* 75.4 (1989), 1119.

72. Barbie Zelizer, "Reading the Past Against the Grain: The Shape of Memory Studies," *Critical Studies in Mass Communication* 12.2 (1995), 216.

73. Burns in Cripps, "Historical Truth," 749.

74. Warren Susman, *Culture as History: The Transformation of American Society in the Twentieth Century* (New York: Pantheon, 1984).

75. David Glassberg, "'Dear Ken Burns': Letters to a Filmmaker," *Mosaic: The Newsletter of the Center on History-Making in America* (at Indiana University), 1 (fall 1991), 3.

76. Burns in Cripps, "Historical Truth," 757.

77. Burns in Thelen, "Movie Maker as Historian," 1033.

78. Burns in Cripps, "Historical Truth," 747.

79. Statistical Research Incorporated, 2.1-2.8.

80. Burns responding to a question after National Press Club newsmaker luncheon speech, "The Making of *The Civil War*," Washington, D.C., 29 October 1990. Available through C-Span Archives (West Lafayette, Indiana), videotape ID# 14743.

81. Burns, interview, 18 February 1993.

82. *The Civil War*, ironically, appeared more tangible and real to many viewers than the closed-circuit video images that were being simultaneously telecast back to America from the Persian Gulf. The nature of living room warfare had become increasingly virtual as high-tech weaponry was now outfitted with cameras which recorded so-called smart bombs hitting an assortment of Iraqi targets, mostly moving vehicles and military installations. Rather than bringing viewers closer to the action, the newer technologies had the odd effect of distancing them by essentially eliminating any sense of the human element altogether.

83. Powers, "Glory, Glory," 218.

84. Ken Burns, "How I Met Lincoln," *American Heritage* 50.4 (1999), 55.

85. Matt Roush, "Ken Burns' Radio Days: His Work is History in the Remaking," *USA Today*, 29 January 1992, 1D.

86. *The Civil War* was not only the most-watched program in public television history, "the audience was [also] the most loyal on record compared to other PBS series," according to Statistical Research Incorporated, 2.2. The average viewer saw 32 percent of the 702 minute duration of this miniseries.

87. Robert Sullivan, "Visions of Glory," *Life*, September 1994, 45.

CHAPTER 2

1. Ken Burns, telephone interview with the author, 18 February 1993.

2. Dayton Duncan, "A Cinematic Storyteller," *Boston Globe Magazine,* 19 March 1989, 77.
3. Robert K. Burns, Jr., "Saint Véran," *National Geographic* 115.4 (1959), 573.
4. Ken Burns quoted in "A Filmmaking Career" on the *Thomas Jefferson* (1997) website at <http://www.pbs.org/jefferson/making/KB_03.htm>.
5. Joyce Wadler, "For Similar Filmmakers, No Civil War, but a Brotherly Indifference," *New York Times,* 17 November 1999, A26.
6. "Ken Burns," *Biography Today* 4.1 (January 1995), 69.
7. Donald D. Jackson, "Ken Burns Puts His Special Spin on the Old Ball Game," *Smithsonian* 25.4 (1994), 40.
8. "Ken Burns and the Historical Narrative on Television," Museum of Television & Radio University Satellite Seminar Series (90 minutes), 19 November 1996.
9. Burns, interview, 18 February 1993.
10. Burns, interview, 18 February 1993.
11. Carroll T. Hartwell, foreword, in *The People, Yes,* photographs and notes by Jerome Liebling (New York: Aperture, 1995), 4.
12. See also Jerome Liebling, *Jerome Liebling Photographs* with essays by Anne Halley and Alan Trachtenberg (Amherst: University of Massachusetts Press, 1982).
13. Burns, interview, 18 February 1993.
14. Ken Burns, "Jerome Liebling as Teacher," text of speech delivered at Hampshire College, 31 March 1990, 3.
15. Jackson, "Ken Burns Puts Special Spin," 41.
16. Morgan Wesson, telephone interview with author, 28 December 1999.
17. Burns, "Jerome Liebling as Teacher," 4.
18. Liebling, *The People, Yes,* 52.
19. "Ken Burns and the Historical Narrative," Museum of Television & Radio University Satellite Seminar Series.
20. Ken Burns quoted in "Interview," *Elizabeth Cady Stanton and Susan B. Anthony* (1999) website at <http://www.pbs.org/stantonanthony/filmmakers/interview.html>.
21. Jackson, "Ken Burns Puts Special Spin," 40.
22. Burns, interview, 18 February 1993.
23. Burns, interview, 18 February 1993.
24. "Ken Burns and the Historical Narrative," Museum of Television & Radio University Satellite Seminar Series.
25. Burns, interview, 18 February 1993.
26. Matt Roush, "Ken Burns' Radio Days: His Work Is History in the Remaking," *USA Today,* 29 January 1992, 1D.
27. "Ken Burns and the Historical Narrative," Museum of Television & Radio University Satellite Seminar Series.
28. Ken Burns in "The Indie Scene: Producer's Interview," September 1998, at <http://www.pbs.org/independents/forum/sept98_forum.2.html>.
29. Duncan, "A Cinematic Storyteller," 77.
30. Burns, interview, 18 February 1993.
31. Tim Clark, "The Man Who Had to Kill Abraham Lincoln," *Yankee* 54 (October 1990), 78.
32. Therese Pasquale-Maguire, "Independent Filmmakers—It's Tough, but Some Are Making It in Massachusetts," *Daily Hampshire Gazette,* 19 May 1981, 4.

33. "Ken Burns and the Historical Narrative," Museum of Television & Radio University Satellite Seminar Series.

34. Tom Lewis, telephone interview with author, 28 December 1999.

35. See Hart Crane, *The Bridge* (New York: Liveright, 1992). This book of poems, originally published in 1930, portrays the Brooklyn Bridge as a mystical unifying symbol for America. Lewis had previously edited *The Letters of Hart Crane and His Family* (New York: Columbia University Press, 1974).

36. See Alan Trachtenberg, *Brooklyn Bridge: Fact and Symbol,* 2nd ed. (Chicago: University of Chicago Press, 1979).

37. David McCullough, *The Great Bridge: The Epic Story of the Building of the Brooklyn Bridge* (New York: Touchstone, 1972).

38. Lewis, interview, 28 December 1999.

39. Tom Lewis, telephone interview with the author, 23 December 1999.

40. Ken Burns's remarks at the "Lewis Mumford at 100" Symposium, Meyerson Hall, University of Pennsylvania, 5 October 1995 (author's audiotaped copy).

41. Burns, "Lewis Mumford at 100" Symposium.

42. "David McCullough," *Current Biography* 54.1 (1993), 387.

43. Burns, interview, 18 February 1993.

44. David McCullough, *The Johnstown Flood: The Incredible Story Behind One of the Most Devastating "Natural" Disasters America Has Ever Known* (reprint, New York: Touchstone, 1987).

45. Karen Heller, "Some People Are Exactly as You Imagine Them, but David McCullough Is Even More So," *Philadelphia Inquirer,* 2 June 1999, 22.

46. Burns, interview, 18 February 1993.

47. The final two-page report submitted to the National Endowment for the Humanities on completing the Brooklyn Bridge Film Project in fulfillment of a $25,000 production grant, dated 27 May 1981, is in the Ken Burns Collection at the Folklore Archives of the Wilson Library, University of North Carolina at Chapel Hill. The *Brooklyn Bridge* materials include all the filmed interviews with various experts and scholars (including outtakes), other footage, test narrations, many copies of letters soliciting financial support, some financial records, various correspondence, along with other related data.

48. Final report to National Endowment for the Humanities, Brooklyn Bridge Film Project, 27 May 1981, 2; Ken Burns Collection.

49. Burns, interview with the author, 27 February 1996; and Ken Burns, "Lewis Mumford at 100" Symposium.

50. Burns, "The Indie Scene: Producer's Interview."

51. Letter from Ken Burns to Eugene J. Bockman (Commissioner, Department of Records, City of New York), 4 December 1980, 2; Ken Burns Collection.

52. Letter from Kenneth Burns to Robin Reynolds (David McCullough's literary agent), 14 March 1980, 2; Ken Burns Collection.

53. "Budget Itemization and Explanation" for *Brooklyn Bridge;* Ken Burns Collection.

54. Lewis, interview, 28 December 1999.

55. Carbon copies of these letters are in the Brooklyn Bridge Film Project materials in the Ken Burns Collection.

56. Burns, interview, 18 February 1993.

57. Robert Sullivan, "Visions of Glory," *Life,* September 1994, 44.

58. Letter from Ken Burns to Martin Beller (program officer at the New York State Council for the Humanities), 15 August 1979; Ken Burns Collection. The $52,400 was comprised of contributions by Abraham & Strauss ($12,500), New York Telephone ($12,500), American Society of Civil Engineers ($12,500), New York State Council for the Arts ($5,000), Consolidated Edison ($5,000), Brooklyn Union Gas ($2,500), Kings County Democratic Committee ($1,000), W.R. Grace & Company ($750), New York Dock Railway ($500), J. Schippers ($100), and Municipal Engineers ($50).

59. The 30 to 1 shooting ratio is determined by the aggregate *Brooklyn Bridge* footage in the Ken Burns Collection.

60. A brief production history is described in the final report submitted to the National Endowment for the Humanities on completing the Brooklyn Bridge Film Project, 27 May 1981; Ken Burns Collection.

61. Letter from Kenneth Burns to Robin Reynolds, 14 March 1980, 2; Ken Burns Collection.

62. Letter from Ken Burns to Eugene J. Bockman, 4 December 1980, 2; Ken Burns Collection.

63. Burns, "The Indie Scene: Producer's Interview."

64. Burns," The Indie Scene: Producer's Interview."

65. Burns, "The Indie Scene: Producer's Interview."

66. Other honors accorded Burns and *Brooklyn Bridge* were a Guggenheim Fellowship, the Christopher Award, the Erik Barnouw Prize from the Organization of American Historians (OAH) for Outstanding Documentary Film in American history, a Blue Ribbon from the American Film Festival, a CINE Golden Eagle, a Special Mention at the Festival Dei Popoli, a First Prize for Documentary at the Birmingham Film Festival, a FILMEX selection, selection for New York's Museum of Modern Art's New Directors series, and selection for the Chicago International Film Festival.

67. Sullivan, "Visions of Glory," 43.

68. Marjorie Rosen and Stephen Sawicki, "Battling Fatigue, Lonely Nights and the Specter of Defeat, Filmmaker Ken Burns Brings *The Civil War* to PBS," *People*, 24 September 1990, 97.

69. Jackson, "Ken Burns Puts Special Spin," 42.

70. Burns, interview, 18 February 1993.

71. Larry Hott, telephone interview with the author, 29 December 1999.

CHAPTER 3

1. Ken Burns, telephone interview with the author, 18 February 1993.

2. Final report to the National Endowment for the Humanities on the Brooklyn Bridge Film Project, 27 May 1981, 2, in the resources on *Brooklyn Bridge* in the Ken Burns Collection at the Folklore Archives of the Wilson Library, University of North Carolina at Chapel Hill.

3. Among miscellaneous journal entries in the resources on *Brooklyn Bridge*, Ken Burns Collection.

4. Leslie Garisto, "Documentarians with a Difference," *New York Times*, 4 August 1985, 21.

5. See Amy Stechler Burns and Ken Burns, *The Shakers—Hands to Work, Hearts to God: The History and Visions of the United Society of Believers in Christ's Second Appearing from 1774 to the Present* with photographs by Ken Burns, Langdon Clay, Jerome Liebling, and from Shaker archives, and foreword by Elderess Bertha Lindsay (New York: Aperture, 1987).

6. Burns and Burns, *The Shakers—Hands to Work, Hearts to God,* 126.

7. Burns and Burns, *The Shakers—Hands to Work, Hearts to God,* 125.

8. Burns and Burns, *The Shakers—Hands to Work, Hearts to God,* 6.

9. "Ken Burns and the Historical Narrative on Television," Museum of Television & Radio University Satellite Seminar Series (90 minutes), 19 November 1996.

10. Ken Burns, interviewed by David Thelen, "The Movie Maker as Historian: Conversations with Ken Burns," *Journal of American History* 81.3 (1994), 1032.

11. Robert S. Root-Bernstein and Michele Root-Bernstein, "Learning to Think With Emotion," *The Chronicle of Higher Education,* 14 January 2000, A64; See also Robert S. Root-Bernstein and Michele Root-Bernstein, *Sparks of Genius: The Thirteen Thinking Tools of the World's Most Creative People* (Boston: Houghton Mifflin, 1999).

12. Ken Burns, interview with the author, 27 February 1996.

13. "Ken Burns and the Historical Narrative," Museum of Television & Radio University Satellite Seminar Series.

14. Ken Burns quoted in a conversation with Mark Gerzon and Molly De Shong entitled "E Pluribus Unum: Ken Burns and the American Dialectic," *Shambhala Sun,* November 1997 at <http://www.shambhalasun.com/Archives/Features/1997/Nov97/KenBurns.htm>.

15. Burns in Thelen, "The Movie Maker as Historian," 1044.

16. Burns in Thelen, "The Movie Maker as Historian," 1046.

17. Larry Hott telephone interview with the author, 29 December 1999.

18. Burns, interview, 18 February 1993.

19. Hott interview, 29 December 1999.

20. Burns, interview, 18 February 1993.

21. Garisto, "Documentarians with a Difference," 22.

22. Burns and Burns, *The Shakers—Hands to Work, Hearts to God,* 120.

23. Geoffrey C. Ward, "Keeping the Shaker Light Burning," *Americana,* July-August 1985, 24. Emphasis added.

24. Garisto, "Documentarians with a Difference," 21.

25. Richard Zoglin, "Feats of Progress," *Time,* 3 February 1992, 55.

26. Garisto, "Documentarians with a Difference," 21.

27. Burns, interview, 18 February 1993.

28. Burns and Burns, *The Shakers—Hands to Work, Hearts to God,* 21.

29. Burns and Burns, *The Shakers—Hands to Work, Hearts to God,* 6.

30. Matt Roush, "Ken Burns' Radio Days: His Work Is History in the Remaking," *USA Today,* 29 January 1992, 1D.

31. Ken Burns in Gerzon and De Shong, "E Pluribus Unum."

32. Zoglin, "Feats of Progress," 55.

33. On resume for Kenneth Lauren Burns, 1991, 4, in the resources on *Baseball* in the Ken Burns Collection at the Folklore Archives of the Wilson Library, University of North Carolina at Chapel Hill.

34. Burns, interview, 27 February 1996.

35. Funding for *The Shakers* was provided by the Ford Foundation, the Independent Documentary Fund of the Corporation for Public Broadcasting, the New York State Council for the Humanities, the New Hampshire Council for the Humanities, the Kentucky Humanities Council, and the Massachusetts Foundation for Humanities and Public Policy.

36. Ken Burns, interviewed by David Jon Wiener, "Interview with 1998 Golden Laurel Recipient: Producer Ken Burns," *Point of View On-Line* 1.1 (1999) at <http://www.empire-pov.com/burns.html>.

37. "Every Man a King: Huey Long and the Rise of Media Politics," grant application submitted to the National Endowment for the Humanities, 26 May 1982.

38. Ken Burns in "The Indie Scene: Producer's Interview," September 1998, at <http://www.pbs.org/independents/forum/sept98_forum.2.html>.

39. "Brooklyn Bridge Film Project," grant application submitted to the National Endowment for the Humanities, 28 August 1978.

40. See Bruce Catton, *The Centennial History of the Civil War*, 3 vols.: *The Coming Fury, Terrible Swift Sword*, and *Never Call Retreat* (Garden City, N.J.: Doubleday, 1961-1965).

41. Geoffrey C. Ward quoted in "Making of the Film—The People—Geoffrey C. Ward/Screenwriter" on the *Thomas Jefferson* (1997) website at <http://www.pbs.org/jefferson/making/ward.htm>.

42. "Ken Burns and the Historical Narrative," Museum of Television & Radio University Satellite Seminar Series.

43. Jack McLaughlin, *Jefferson and Monticello: The Biography of a Builder* (New York: Henry Holt, 1988), 12.

44. Ken Burns in response to a question during the Historians Film Committee panel, "Television and History: Ken Burns's *Thomas Jefferson*," at the 112th Annual Meeting of the American Historical Association, 10 January 1998, Seattle, Washington.

45. Letter from Richard F. Snow to Ken Burns, 4 November 1985, 2, in the resources on *The Civil War* in the Ken Burns Collection at the Folklore Archives of the Wilson Library, University of North Carolina at Chapel Hill.

46. Garisto, "Documentarians with a Difference," 21.

47. Joel Sternberg, "David L. Wolper, U.S. Producer," in *Museum of Broadcast Communications Encyclopedia of Television*, vol. 3, ed. Horace Newcomb (Chicago: Fitzroy Dearborn, 1997).

48. "General Treatment" (this specific quotation is on the first of two pages), in the Ken Burns Collection at the Folklore Archives of the Wilson Library, University of North Carolina at Chapel Hill. *The Statue of Liberty* materials include all drafts of the script, all filmed interviews with transcripts, other footage, cue sheets, other related data.

49. "General Treatment," 2, in the resources on *The Statue of Liberty* in the Ken Burns Collection.

50. Paul Barnes quoted in "Making of the Film—The People—Paul Barnes/Editor" on the *Thomas Jefferson* (1997) website at <http://www.pbs.org/jefferson/making/barnes.htm>.

51. Ken Burns, interviewed by Thomas Cripps, "Historical Truth: An Interview with Ken Burns," *American Historical Review* 100.3 (1995), 749.

52. Robert Penn Warren, *All the King's Men* (New York: Harvest, 1996).

53. Burns, interview, 18 February 1993.

54. Burns, interview, 18 February 1993.

55. Ken Burns's written reactions appended to the cover of Geoffrey C. Ward's first draft of the *Huey Long* script, 4 July 1984, in the Ken Burns Collection at the Folklore Archives of the Wilson Library, University of North Carolina at Chapel Hill. The *Huey Long* materials include all drafts of the script, all filmed interviews with transcripts, other footage, production notes, timing sheets, various correspondence, other related data.

56. Letter from William E. Leuchtenburg to Ken Burns, 19 September 1984, 2, in the resources on *Huey Long*, Ken Burns Collection.

57. Larry Leventhal, "One Man's 'Civil War' Is Another's Foundation," *Variety*, 21 September 1992, 80.

58. See "Huey Pierce Long, 1928-1932," on the Louisiana Secretary of State/Louisiana Governor website at <http://www.sec.state.la.us/60.htm>.

59. Burns, interview, 18 February 1993.

60. Warren, *All the King's Men*, 427.

61. John C. Tibbetts, "The Incredible Stillness of Being: Motionless Pictures in the Films of Ken Burns," *American Studies* 37.1 (Spring 1996), 117-33. Tibbetts provides the most thorough examination of Burns's still-in-motion technique to date.

62. Susan Sontag, *On Photography* (New York: Doubleday, 1978), 71.

63. "Ken Burns and the Historical Narrative," Museum of Television & Radio University Satellite Seminar Series.

64. Vincent Canby, "'Huey Long,' a Documentary on the Louisiana Populist, by Ken Burns," *New York Times*, 28 September 1985, 9.

65. Ken Bode, "Hero or Demagogue? The Two Faces of Huey Long on Film," *New Republic*, 3 March 1986, 37.

66. Jim Wooten, "His Bullhorn Demagoguery Might Have Been Too Much for TV," *TV Guide*, 11 October 1986, 15.

67. "Ken Burns," *Current Biography*, May 1992, 7.

68. Dayton Duncan, "A Cinematic Storyteller," *Boston Globe Magazine*, 19 March 1989, 80.

69. Duncan, "A Cinematic Storyteller," 82.

70. "*The Civil War:* A Television Series," grant application with appendices submitted to the National Endowment for the Humanities, 14 March 1986, 68.

71. Ken Burns, "Sharing the American Experience," text of a speech delivered at the Norfolk Forum, Chrysler Auditorium, Norfolk, Virginia, 27 February 1996, 11.

72. Leventhal, "One Man's 'Civil War' Is Another's Foundation," 80.

73. "*Thomas Hart Benton:* An American Original," grant application submitted to the National Endowment for the Humanities, 12 March 1986, 8.

74. "*Thomas Hart Benton:* An American Original," 8.

75. "*Thomas Hart Benton:* An American Original," 8.

76. Burns actually decided on this strategy later in the assembly process as he had both actor Jason Robards, the narrator, and Mike Wallace, the CBS newsman who was a neighbor of Benton's in the summers on Martha's Vineyard, read versions of the introduction before going with the three graphics and no voice-over. Scripts and audiotapes of Robards's and Wallace's readings are in the Ken Burns Collection at the Folklore Archives of the Wilson Library, University of North Carolina at Chapel Hill. The *Thomas Hart Benton* materials include some drafts of the script, all the filmed

interviews with transcripts, other footage, film logs, various correspondence, other related data.

77. Valerie Lester, "Happy Birthday, Tom Benton!" *Humanities* 10.6 (1989), 32.

78. "Ken Burns and the Historical Narrative," Museum of Television & Radio University Satellite Seminar Series.

79. "*Thomas Hart Benton:* An American Original," 43.

80. Bank statements indicate that WETA made an initial deposit of $200,000 out of a $661,471 total budget on September 2, 1986, thus formally setting the preparation and research process into motion. These financial records are available in the Ken Burns Collection at the Folklore Archives of the Wilson Library, University of North Carolina at Chapel Hill. *The Congress* materials include all drafts of the script, all filmed interviews with transcripts, other footage, filming schedules, logs, financial statements, various correspondence, other related data.

81. Duncan, "A Cinematic Storyteller," 80.

82. The 12 legislators featured in "The Progressives" sequence are Joseph Gurney ("Uncle Joe") Cannon of Illinois, George W. Norris of Nebraska, Robert M. La Follette of Wisconsin, Janette Rankin of Montana, Nicholas Longworth of Ohio, Fiorello H. LaGuardia of New York, Robert F. Wagner of New York, Sam Rayburn of Texas, Hugo L. Black of Alabama, Alben Barkley of Kentucky, Harry S. Truman of Missouri, and Robert M. La Follette, Jr. of Wisconsin.

83. Ken Burns in Gerzon and De Shong, "E Pluribus Unum."

84. "Ken Burns," *Current Biography,* May 1992, 7.

85. Ken Burns in response to a question during the Historians Film Committee panel, "Television and History: Ken Burns's *Thomas Jefferson.*"

86. Ken Burns in Wiener, "Interview with 1998 Golden Laurel Recipient."

CHAPTER 4

1. "Ken Burns and the Historical Narrative on Television," Museum of Television & Radio University Satellite Seminar Series (90 minutes), 19 November 1996. *Empire of the Air* was Ken Burns's eighth major PBS special. He also supervised and coproduced *Lindbergh,* which was produced and directed by Stephen Ives.

2. Thomas S. W. Lewis, "Radio Revolutionary: Edwin Howard Armstrong's Invention of FM Radio," *American Heritage of Invention & Technology,* Fall 1985, 34-41.

3. Tom Lewis, telephone interview with the author, 23 December 1999.

4. Lewis, interview, 23 December 1999.

5. Tom Lewis, telephone interview with the author, 28 December 1999.

6. Ken Burns, telephone interview with the author, 18 February 1993.

7. Tom Lewis, *Empire of the Air: The Men Who Made Radio* (New York: HarperPerennial, 1993), 401.

8. The chief archives are Armstrong's papers at the Armstrong Memorial Research Foundation and the Butler Library at Columbia University; Lee de Forest's papers at the Library of Congress; David Sarnoff's papers at the David Sarnoff Research Center in Princeton, New Jersey; the George H. Clark Collection of Radioana at the National Museum of American History, Smithsonian Institution; the Antique Wireless Museum

in Holcomb, New York; the Museum of Broadcasting (now the Museum of Television & Radio) in New York; and the Broadcast Pioneers Library in Washington, D.C.

9. See Erik Barnouw, *A Tower in Babel: A History of Broadcasting in the United States to 1933* (New York: Oxford, 1966); Erik Barnouw, *The Golden Web: A History of Broadcasting in the United States, 1933-1953* (New York: Oxford, 1968); Erik Barnouw, *The Image Empire: A History of Broadcasting in the United States from 1953* (New York: Oxford, 1970); Kenneth Bilby, *The General: David Sarnoff and the Rise of the Communications Industry* (New York: Harper & Row, 1986); and Susan J. Douglas, *Inventing American Broadcasting, 1899-1922* (Baltimore: Johns Hopkins University Press, 1987).

10. The 25 people interviewed for *Empire of the Air* fall broadly into four categories: scholars—Erik Barnouw and Susan Douglas; radio engineers (many of whom knew Armstrong, de Forest, and/or Sarnoff)—Harold Beveridge, Tom Buzalski, Frank Gunther, Loren Jones, Robert Morris, and Bruce Roloson; witnesses (personal acquaintances of Armstrong, de Forest, or Sarnoff)—Kenneth Bilby, Jeanne Hammond (Armstrong's niece), Charlotte Katchurin (Sarnoff's secretary), Eleanor Peck (de Forest's daughter), Dana Raymond (lawyer and friend of Armstrong and his wife, Marion), Josephine Raymond (friend of the Armstrongs), Robert Sarnoff (son), Gertrude Tyne (friend of de Forest); and radio personalities, broadcasters, amateurs, or archivists—Red Barber, John Beach, Norman Corwin, Garrison Keillor, Bruce Kelley, Helen Kelley, Maynard Edmeston, Robert Saudek, and Eric Sevareid.

11. "Radio Pioneers" grant application submitted to the National Endowment for the Humanities, 17 March 1989.

12. "Radio Pioneers," NEH grant application, cover materials.

13. Ken Fink went on to an active career directing prime-time hour-long dramas and TV movies, most notably for *Homicide: Life on the Streets* (NBC, 1993-1999) and *The Vernon Johns Story: The Road to Freedom* (1994), starring James Earl Jones.

14. Morgan Wesson, telephone interview with the author, 28 December 1999.

15. Lewis, interview, 28 December 1999.

16. Wesson, interview, 28 December 1999.

17. "Ken Burns Looks at Broadcasting," *Broadcasting*, 5 August 1991, 28.

18. Wesson, interview, 28 December 1999.

19. Lewis, interview, 23 December 1999.

20. VV/RP Corwin, 1, and SR 38 (filmed interview) in the Ken Burns Collection at the Folklore Archives of the Wilson Library, University of North Carolina at Chapel Hill. The *Empire of the Air* materials include all five drafts of the script (dated 12/13/90, 1/5/91, 3/6/91, 4/11/91, and 6/17/91, respectively), all the filmed interviews with various witnesses and scholars (including outtakes), other footage, notes on decision-making, test narrations, some financial records, correspondence, other related data.

21. Burns, interview, 18 February 1993.

22. Wesson, interview, 23 December 1999.

23. Lewis, interview, 23 December 1999.

24. Wesson, interview, 23 December 1999.

25. Lewis, interview, 28 December 1999.

26. "Radio Pioneers" script 12/13/90, box 28, Ken Burns Collection.

27. A rough draft of the book manuscript including comments and production notes penned into the margins is contained in box 28 of the Ken Burns Collection. Ward,

Burns, and Barnes were evidently referring to this manuscript throughout the production process.

28. Electrical engineer Edwin Howard Armstrong invented three of the most fundamental discoveries underlying all modern radio, radar, and television. His 1912 regenerative circuit vastly improved de Forest's audion by enhancing radio-wave detection and amplification to an extraordinary degree; it caused the audion to oscillate rapidly and, therefore, behave like a transmitter as well as a more powerful and selective receiver. The regenerative circuit basically resulted in the first radio amplifier and, additionally, produced continuous-wave transmissions which created the basis for all later radio broadcasting. His 1918 superheterodyne circuit adapted a technique called heterodyning found in early wireless to produce unsurpassed amplification that no longer required headphones, thus rendering crystal sets obsolete. The "superhet" also made tuning far easier and more precise. His 1933 invention of FM (frequency modulation) was a revolutionary new system of radio which solved the chronic noise and static problem of AM (amplitude modulation). FM offered a much clearer signal and a far higher fidelity sound than was imaginable in radio broadcasting up to that point.

29. Hayden White, *Tropics of Discourse: Essays in Cultural Criticism* (Baltimore: Johns Hopkins University Press, 1978), 50.

30. Burns, interview, 18 February 1993.

31. Wireless telegraphy is the transmission of written matter through the air from point-to-point by a signal code (e.g., Morse code). Wireless telephony is the transmission of the voice or other sounds (e.g., music) through the air from point-to-point. Radio broadcasting is multipoint radio telephony.

32. Lewis, *Empire of the Air*, 5.

33. John Budris, "When America Stopped to Listen," *Christian Science Monitor*, 28 January, 1992, 14.

34. Lewis, *Empire of the Air*, 5.

35. "Radio Pioneers," NEH grant application, 5.

36. TT/RP Keillor, 7, and SR 36/37 (filmed interview), Ken Burns Collection.

37. Ken Burns, "Notes On Viewing Rough Cut" (1/13/91–1/14/91) and "Radio Pioneers" Scripts 12/13/90 and 1/5/91, box 28, Ken Burns Collection.

38. Wesson, interview, 28 December 1999.

39. See Amy Henderson, *On the Air: Pioneers of American Broadcasting* (Washington, D.C.: Smithsonian Institution Press, 1988), 24-43; Museum of Television & Radio (formerly Museum of Broadcasting) opened its doors on Manhattan's East 53rd Street in 1976. Today, literally tens of thousands of people visit this repository weekly to view and listen to excerpts from more than 50 thousand hours of broadcast and cable programming.

40. Harry F. Waters, "A Less-than-Civil War," *Newsweek*, 27 January 1992, 61.

41. Burns, interview, 27 February 1996.

42. Erik Barnouw, ed., *International Encyclopedia of Communications*, 4 vols. (New York: Oxford, 1989).

43. Erik Barnouw, interview with the author, 18 January 2000.

44. Lewis, interview, 28 December 1999.

45. "Radio Pioneers," NEH grant application, 41.

46. "Radio Pioneers," NEH grant application, 6.

47. QQ/RP Douglas, 1-24, and SR 29/30/31 (filmed interview), Ken Burns Collection.

48. Susan Douglas, telephone interview with the author, 24 January 2000.
49. "Radio Pioneers" scripts dated 12/13/90, 1/5/91, 3/6/91, 4/11/91, and 6/17/91; notes on various rough cuts in box 28, Ken Burns Collection.
50. Burns, interview, 18 February 1993.
51. Budris, "When America Stopped to Listen," 14.
52. Filmed interviews in the Ken Burns Collection: AB/RP Beveridge, 2, and SR 41; NN/RP Roloson, 1, and SR 24/25/26; AC/RP Jones, 5, and SR 42; HH/RP Buzalski, 4, and SR 15; RR/RP Tyne, 2, and SR 32/33; AE/RP Peck, 1, and SR 47; DD/RP Raymond (Josephine), 2, and SR 7/8; CC/RP Raymond (Dana), 1, and SR 7; SS/RP Hammond, 18, and SR 34/35; EE/RP Morris, 5, and SR 9/10/11; AC/RP Jones, 4, and SR 42; HH/RP Buzalski, 1, and SR 15; HH/RP Buzalski, 4, and SR 15; BB/RP Bilby, 10, and SR 5/6; XX/RP Katchurin, 4, and SR 39; AC/RP Jones, 5, and SR 42; ZZ/RP Sarnoff, 4, and SR 40.
53. See Ken Burns's notes on interview transcripts AA through ZZ; "Radio Pioneers" scripts dated 12/13/90, 1/5/91, 3/6/91, 4/11/91, and 6/17/91; and various rough cuts in box 28, Ken Burns Collection.
54. AA/RP Barnouw, 4, and SR 2/3/4 (filmed interview), Ken Burns Collection.
55. QQ/RP Douglas, 7, and SR 29/30/31 (filmed interview), Ken Burns Collection.
56. BB/RP Bilby, 1, 4, 11, and SR 5/6 (filmed interview), Ken Burns Collection. See also Bilby, *The General: David Sarnoff*.
57. AC/RP Jones, 9, and SR 42 (filmed interview), Ken Burns Collection.
58. See Lee de Forest, *Father of Radio: The Autobiography of Lee de Forest* (Chicago: Wilcox & Follet, 1950), 4. In this autobiography, de Forest writes with characteristic romance and hyperbole: "Unwittingly then I had discovered an Invisible Empire of the Air, intangible, yet solid as granite, whose structure shall persist while man inhabits the planet; a global organism, imponderable yet most substantial, both mundane and empyreal; fading not as the years, the centuries fade away—and electronic fabric influencing all our thinking, making our living more noble. For this, my life has been rich indeed!"
59. Barnouw, interview, 18 January 2000.
60. Lewis, interview, 28 December 1999.
61. See Susan J. Douglas, *Listening In: Radio and the American Imagination, from Amos n' Andy and Edward R. Murrow to Wolfman Jack and Howard Stern* (New York: Times Books, 1999); Michele Hilmes, *Radio Voices: American Broadcasting, 1922-1952* (Minneapolis: University of Minnesota Press, 1997); J. Fred MacDonald, *Don't Touch That Dial!: Radio Programming in American Life from 1920 to 1960* (Chicago: Nelson-Hall, 1979); and Barbara Dianne Savage, *Broadcasting Freedom: Radio, War, and the Politics of Race, 1938-1948* (Chapel Hill: University of North Carolina Press, 1999).
62. Jib Fowles, "Three Men Who Truly Made Radio: Ken Burns' Empire of the Air: The Men Who Made Radio (PBS)," *Historical Journal of Film, Radio & Television* 14.2 (1994), 219.
63. Wesson, interview, 28 December 1999.
64. Steven O. Shields, "Book Review of Tom Lewis's *Empire of the Air: The Men Who Made Radio*," *Journal of Radio Studies* 1.1 (1992), 177.
65. Jib Fowles, "Three Men Who Truly Made Radio," 219. Fowles makes a strong case in his review for "the magnitude of Marconi's place in radio history," citing not only his ingenious and workable system of wireless telegraphy but his long-term business

acumen in successfully marketing and selling his electronics products. He also acknowledges Reginald Fessenden for inventing wireless telephony or over-the-air voice transmission and reception. And finally, he identifies the "desires of the mass audience" as the deciding factor in influencing the "evol[ution of] wireless . . . into radio."

66. Jim Cullen, "Review of *Empire of the Air: The Men Who Made Radio* by Ken Burns, Morgan Wesson, and Tom Lewis," *Journal of American History* 78.4 (1992), 1290. See Daniel J. Czitrom, *Media and the American Mind: From Morse to McLuhan* (Chapel Hill: University of North Carolina Press, 1982) and Susan Smulyan, *Selling Radio: The Commercialization of American Broadcasting, 1920-1934* (Washington, D.C.: Smithsonian Institution Press, 1994).

67. Lewis, interview, 28 December 1999.

68. Tom Lewis, *Empire of the Air: The Men Who Made Radio* (New York: Edward Burlingame, 1991).

69. David Ossman, Phil Austin, Phil Proctor, and Peter Bergman founded the Firesign Theatre in 1966 on Radio Free Oz in Los Angeles. They created 14 critically acclaimed and popular comedy albums together. Ossman left the troupe in 1985 to pursue other independent projects in radio theater.

70. Richard Zoglin, "White Men Behaving Badly," *Time,* 16 September 1996, 77.

71. David Ossman directed *War of the Worlds 50th Anniversary Production* in 1988. This effort was based on Howard Koch's original script and Orson Welles's production and direction. The updated version starred Jason Robards and Steve Allen among others and was recorded at George Lucas's Skywalker Ranch studios, employing the latest technology to simulate a contemporary audio experience. Ossman also directed Norman Corwin's critically acclaimed 1991 production of *We Hold These Truths* for American Public Radio.

72. The radio adaptation of *Empire of the Air: The Men Who Made Radio* is available through LodesTone Audio Theatre, which is the most comprehensive source in the United States for new works in audio theater. LodesTone can be reached by writing to LodesTone Catalog, 611 Empire Mills Road, Bloomington, Indiana 47401.

73. Lewis, interview, 28 December 1999.

74. Seth Goldstein, "Whole New Ball Game," *Billboard,* 23 April 1994, 77.

75. Eileen Fitzpatrick, "Burns Sues Pacific Arts for Back Royalties on Videos," *Billboard,* 22 October 1994, 8.

76. The other historical documentaries now packaged with *Empire of the Air: The Men Who Made Radio* by Turner Home Entertainment are *Brooklyn Bridge, The Shakers: Hands to Work, Hearts to God, The Statue of Liberty, Huey Long, Thomas Hart Benton, The Congress.*

77. Trudi Miller Rosenblum, "History Repeats Itself on Video: Ken Burns's Series Have Created a Mainstream Genre," *Billboard,* 13 January 1996, 56.

CHAPTER 5

1. Ken Burns, interviewed by Erik Spanberg, "Commentary: Ken Burns on *Civil War, Clinton, Success," Business Journal of Charlotte,* 1 March 1999, at <http://www.amcity.com:80/charlotte/stories/1999/03/01/editorial3.html>.

2. Albert Kim, "The Filmmaker Behind PBS' 'Civil War' Steps Up to the Plate with His Epic 'Baseball,'" *Entertainment Weekly,* 16 September 1994, 41.

3. "*Baseball:* A Documentary Film Series for Public Television," grant application submitted to the National Endowment for the Humanities, 7 March 1991, 4.

4. "*Baseball:* A Documentary," NEH grant application, 4.

5. "*Baseball:* A Documentary," NEH grant application, 5.

6. "*Baseball:* A Documentary," NEH grant application, cover sheet.

7. Lynn Novick, "Progress Report on the Baseball Film Project," submitted to the National Endowment for the Humanities, 28 February 1994, 2, in the resources on *Baseball* in the Ken Burns Collection at the Folklore Archives of the Wilson Library, University of North Carolina at Chapel Hill. The *Baseball* materials include all drafts of the script, all the filmed interviews with various scholars and experts (including outtakes), other footage, notes on decision making, test narrations, some financial records, correspondence, other related data.

8. Lynn Novick, "Progress Report on the Baseball Film Project," submitted to the National Endowment for the Humanities, 28 February 1994, 2, in the resources on *Baseball,* Ken Burns Collection.

9. "Florentine Films Budget Report," 7 April 1994, 2 pages, in the resources on *Baseball,* Ken Burns Collection. B. J. Bullert, *Public Television: Politics and the Battle over Documentary Film* (New Brunswick, N.J.: Rutgers University Press, 1997) reports on page 224: "With *Baseball,* CPB and PBS contributed about $2 million; General Motors gave $2.1 million toward production; the National Endowment for the Humanities came up with $2 million; the Pew Charitable Trusts gave $987,000, and the Arthur Vining Davis Foundation donated $550,000."; Kim, "Filmmaker Steps Up to the Plate," 41.

10. Tom Shales, "'Civil War': A Triumph on All Fronts," *Washington Post,* 20 September 1990, E4.

11. James Day, *The Vanishing Vision: The Inside Story of Public Television* (Berkeley: University of California Press, 1995), 314-15.

12. Day, *Vanishing Vision,* 315.

13. Michael Walker, "Covering All the Bases," *Entertainment Weekly,* 9 September 1994, 12.

14. Robert N. Gold, Esq., "Memorandum—The History Company," 3 August 1993, in the resources on *Baseball,* Ken Burns Collection. See also John Thorn, Pete Palmer, Michael Gershman, and David Pietrusza, eds., *Total Baseball: The Ultimate Encyclopedia of Baseball,* 6th ed. (New York: Total Sports, 1999).

15. Gold, "Memorandum—The History Company," 3, in the resources on *Baseball,* Ken Burns Collection.

16. Carbone, Smolan, and Shade Associates, "Products for Baseball," in the resources on *Baseball,* Ken Burns Collection.

17. Larry Leventhal, "One Man's 'Civil War' Is Another's Foundation," *Variety,* 21 September 1992, 81.

18. Leventhal, "One Man's 'Civil War,'" 81.

19. Bullert, *Public Television,* 179.

20. Bullert, *Public Television,* 179.

21. A partial listing of these various events is given in Owen Comora Associates, "'Baseball' Personal Appearances, 26 May 1994, in the resources on *Baseball*, Ken Burns Collection.

22. GM/*Baseball*, "GM Playbook Overview," 12 May 1994, 8, in the resources on *Baseball*, Ken Burns Collection.

23. Todd Brewster, "*Baseball* Extends Its Reach," *Baseball: A Film By Ken Burns* (promotional newsletter) 1.3, June 1994, 4,

24. Michael McCarthy, "GM Aboard with Burnsball," *Brandweek*, 1 August 1994, 9.

25. Chris Morris and Eileen Fitzpatrick, "Firms Go to Bat for 'Baseball,'" *Billboard*, 13 August 1994, 1.

26. See Geoffrey C. Ward and Ken Burns, *Baseball: An Illustrated History* (New York: Knopf, 1994); Geoffrey C. Ward, Ken Burns, and S. A. Kramer, *25 Great Moments (Baseball, the American Epic)* (New York: Random House, 1994); Geoffrey C. Ward, Ken Burns, and Jim O'Connor, *Shadow Ball: The History of the Negro Leagues (Baseball, the American Epic)* (New York: Random House, 1994); and Geoffrey C. Ward, Ken Burns, and Paul Robert Walker, *Who Invented the Game (Baseball, the American Epic)* (New York: Random House, 1994).

27. Morris and Fitzpatrick, "Firms Go to Bat,'"1.

28. Jonathan Yardley, "A Showcase of 'Baseball'?" *Washington Post*, 11 July 1994, B2.

29. Kim, "Filmmaker Steps Up to the Plate," 41.

30. Robert Sullivan, "Visions of Glory," *Life*, September 1994, 41.

31. Letter from Julie Dunfey to Curt Flood, 9 May 1994, 1, in the resources on *Baseball*, Ken Burns Collection.

32. Letter from Gerard F. McCauley (Burns's literary agent) to Lynn Novick, 3 May 1994, 1, in the resources on *Baseball*, Ken Burns Collection; Walker, "Covering All the Bases," 12.

33. Walker, "Covering All the Bases," 12.

34. Ken Burns, interview with the author, 27 February 1996.

35. Sullivan, "Visions of Glory," 41.

36. Jack Ketch, "Touching All the Bases," *The World and I*, January 1995, 151; Trudi Miller Rosenblum, "History Repeats Itself on Video," *Billboard*, 13 January 1996, 56.

37. Elaine Santaro, "Non-Profits Win with *Baseball* Series," *Fund Raising Management* 25.9 (November 1994), 10.

38. Santaro, "Non-Profits Win with *Baseball* Series," 11.

39. Walker, "Covering All the Bases," 12.

40. "Ken Burns," *Current Biography*, May 1992, 10.

41. Harry F. Waters, "Talkin' *Baseball:* Interview (with Ken Burns)," *Newsweek*, 12 September 1994, 67; "Ken Burns," *Biography Today*, 1995 Annual Cumulation (Detroit: Omnigraphics, 1995), 70.

42. Sullivan, "Visions of Glory," 41.

43. "Historian Goodwin on Baseball Memoir," *Charlie Rose Show* (#2063), airdate: 30 December 1997.

44. Ken Burns, interview by David Thelen, "The Movie Maker as Historian: Conversations with Ken Burns," *Journal of American History* 81.3 (1994), 1044.

45. Rod Beaton, "Documentary Offers Historical Tidbits, Modern-Day Insights," *USA Today*, 25 August 1994, 3C.

46. Geoffrey C. Ward, "*Baseball* Outline—Red Sox/Dodgers," 1990, 2, in the resources on *Baseball,* Ken Burns Collection.

47. See Kenneth S. Robson, ed., *A Great and Glorious Game: Baseball Writings of A. Bartlett Giamatti,* with a foreword by David Halberstam (New York: Algonquin, 1998), 7. This quote is found in the first paragraph of the often-cited lyrical essay "The Green Fields of the Mind."

48. Ward, "*Baseball* Outline—Red Sox/Dodgers," 2, in the resources on *Baseball,* Ken Burns Collection.

49. Ward, "*Baseball* Outline—Red Sox/Dodgers," 3, in the resources on *Baseball,* Ken Burns Collection.

50. "Ken Burns and 'Baseball' Talk," *Charlie Rose Show* (#1211), airdate: 23 September 1994.

51. Ward and Burns, *Baseball: An Illustrated History,* 198.

52. Letter from Deborah A. Shattuck, Capt. U.S.A.F., to Lynn Novick, 10 February 1991, 1, in the resources on *Baseball,* Ken Burns Collection.

53. Letter from William E. Leuchtenberg to Ken Burns, 10 March 1993, 2, in the resources on *Baseball,* Ken Burns Collection.

54. Letter from Roger Angell to Ken Burns and Geoff Ward, 19 February 1991, 1, in the resources on *Baseball,* Ken Burns Collection.

55. Letter from John Holway to Lynn Novick, 4 March 1991, 1, in the resources on *Baseball,* Ken Burns Collection.

56. See "The Making of *Baseball*" videotape (1994); "Inning 5—General Comments," in the resources on *Baseball,* Ken Burns Collection.

57. Gerald Early, "American Integration, Black Heroism, and the Meaning of Jackie Robinson," *Chronicle of Higher Education,* 23 May 1997, B4.

58. Ken Burns telephone interview with the author, 18 February 1993.

59. "*Baseball:* A Documentary," NEH grant application, 109.

60. Ward and Burns, *Baseball: An Illustrated History,* 292.

61. George Weigel, "Politically-Correct Baseball," *Commentary,* November 1994, 48.

62. Donald Dale Jackson, "Ken Burns Puts His Special Spin on the Old Ball Game," *Smithsonian* 25.4 (1994), 40.

63. Ken Burns responding to a question after delivering the speech, "The Making of *Baseball*," at a National Press Club newsmaker luncheon, 12 September 1994 (Washington, D.C.: C-Span, 1994), 10.

64. Burns in Thelen, "The Movie Maker as Historian," 1044.

65. Frank McConnell, "No Fall Classic: Burns's 'Baseball,'" *Commonweal,* 18 November 1994, 31.

66. McConnell, "No Fall Classic,'" 32.

67. Harry F. Waters, "Baseball Is Forever," *Newsweek,* 12 September 1994, 66, 68.

68. Richard Zoglin, "White Men Behaving Badly," *Time,* 16 September 1996, 77.

69. Robert Sullivan, "The Burns Method: History or Myth-Making?" *Life,* September 1994, 42.

70. Charles Krauthammer, "Requiem for the Summer Game," *Time,* 3 April 2000, 100.

71. Ken Burns quoted in "The Making of 'Not For Ourselves Alone,'" PBS promotional special, airdate: 7 November 1999.

72. Sullivan, "Visions of Glory," 45.

73. Dichter quoted in *The Making of Baseball* (1994).

74. Jerry Adler, "Revisiting the Civil War," *Newsweek,* 8 October 1990, 59

75. Jackson, "Ken Burns Puts His Special Spin," 42.

76. Jackson, "Ken Burns Puts His Special Spin," 42.

77. Paul Barnes quoted in "The Making of 'Not For Ourselves Alone,'" PBS promotional special, airdate: 7 November 1999.

78. Ken Burns, opening remarks on the Historians Film Committee panel "Television and History: Ken Burns's *Thomas Jefferson,*" at the 112th Annual Meeting of the American Historical Association, 10 January 1998, Seattle, Washington.

79. Jackson, "Ken Burns Puts His Special Spin," 42.

80. Ken Burns quoted in "A Conversation with Filmmaker Ken Burns," Part II, PBS promotional special, airdate: 5 November 1997.

81. "*The West:* A Documentary Film Series for Public Television," grant application submitted to the National Endowment for the Humanities, 2 September 1992, 20.

82. "*The West:* A Documentary Film Series," NEH grant application, 20.

83. "*The West:* A Documentary Film Series," NEH grant application, 20.

84. "*The West:* A Documentary Film Series," NEH grant application, 21.

85. "'The West'—Ken Burns Documentary," *Charlie Rose Show* (#1723), airdate: 10 September 1996.

86. Lawrie Mifflin, "A Film Maker Crosses a Frontier of His Own," *New York Times,* 8 July 1996, C10.

87. "Stephen Ives: Intersecting History," *The West Press Kit* (New York: Owen Comora Associates, a Division of Serino Coyne Public Relations, 1996), 2.

88. Mifflin, "A Film Maker Crosses a Frontier of His Own," C10.

89. Karen Everhart Bedford, "*The West:* Stephen Ives Unreels a Complex Saga of Glory, Shame and All the Rest," *Current,* 17 June 1996, at <http://www.current.org/hi/hi611w.html>.

90. "*The West:* A Documentary Film Series," NEH grant application, 144-145.

91. "*The West:* A Documentary Film Series," NEH grant application, 21.

92. Gregory Lalire, "Interview: *The West* According to Burns and Ives," *Wild West,* October 1996, at <http://www.thehistorynet.com/WildWest/articles/1096_text.htm>.

93. Joseph Flanagan, "The Other Burns Comes Back East: With *The Way West,*" *Current,* 6 June 1994, at <http://www.current.org/hi411.html>.

94. Joyce Wadler, "For Similar Filmmakers, No Civil War, but a Brotherly Indifference," *New York Times,* 17 November 1999, A26.

95. John F. Kasson, *Amusing the Millions: Coney Island at the Turn of the Century* (New York: Hill & Wang, 1978).

96. Ric Burns quoted in "The Ric Burns Interview" on *The Donner Party* (1992) website at <http://www.pbs.org/wgbh/amex/donner/ricburns.html>. Emphasis added.

97. Karen Everhart Bedford and Steve Behrens, "With Funding as Shaky as Ever, the Craft of Historical Documentaries Hits New Highs," *Current,* 20 January 1997, at <http://www.current.org/hi/hi701.html>.

98. Alan Brinkley, "The Western Historians: Don't Fence Them In," *New York Times Book Review,* 20 September 1992, 27.

99. Thomas Schatz, *Hollywood Genres: Formulas, Filmmaking, and the Studio System* (Philadelphia: Temple University Press, 1981), 47-48. Emphasis added.

100. John G. Cawelti, *The Six-Gun Mystique,* 2nd ed. (Bowling Green, Ohio: Bowling Green State University Press, 1984), 2.

101. Cawelti, *Six-Gun Mystique,* 16.

102. Cawelti, *Six-Gun Mystique,* 16-20.

103. Cawelti, *Six-Gun Mystique,* 20-21.

104. Rick Marin, "The Other 'Civil War'—TV: 'The Way West' is a Tragedy Movingly Retold," *Newsweek,* 8 May 1995, 62; Richard Zoglin and Martha Smilgis, "Back from Boot Hill," *Time,* 15 November 1993, 90.

105. Marin, "The Other 'Civil War'—TV," 62.

106. John G. Neihardt, *Black Elk Speaks: Being the Life Story of a Holy Man of the Ogalala Sioux* (Lincoln: University of Nebraska Press, 1988).

107. "'The Way West'—Ric Burns," *Charlie Rose Show* (#1371), airdate: 5 May 1995.

108. Bedford, " *The West:* Stephen Ives Unreels a Complex Saga."

109. Mifflin, "A Film Maker Crosses a Frontier of His Own," C9.

110. Ed Bark, "Ken and Ric Burns Are Close, Relatively Speaking," *Dallas Morning News,* 3 August 1999, 16.

111. Flanagan, "The Other Burns Comes Back East: With *The Way West.*"

112. Bedford, " *The West:* Stephen Ives Unreels a Complex Saga."

113. "Ken Burns: Citizen of the West," 2, *The West Press Kit.*

114. Geoffrey C. Ward, *The West: An Illustrated History,* with a preface by Stephen Ives and Ken Burns (Boston: Little, Brown, 1996), 434.

115. Bedford, " *The West:* Stephen Ives Unreels a Complex Saga."

116. Letter from Ramón A. Gutiérrez to Stephen Ives, 17 August 1992, 1, in the resources on *Baseball* (containing a folder on *The West*), Ken Burns Collection.

117. Bedford, " *The West:* Stephen Ives Unreels a Complex Saga."

118. Mifflin, "A Film Maker Crosses a Frontier of His Own," C10.

119. Bedford, " *The West:* Stephen Ives Unreels a Complex Saga."

120. " *The West:* A Documentary Film Series," NEH grant application, 65.

121. Lalire, " *The West* According to Burns and Ives."

122. See J. S. Holliday, *The World Rushed In: The California Gold Rush Experience,* reprint edition (New York: Simon & Schuster, 1983).

123. Lalire, " *The West* According to Burns and Ives."

124. Lalire, " *The West* According to Burns and Ives."

125. "N. Scott Momaday: Keeper of the Flame," 2, *The West Press Kit.*

126. Gutiérrez to Ives, 17 August 1992, 3.

127. "Stephen Ives: Intersecting History," 2, *The West Press Kit.*

128. Sullivan, "Visions of Glory," 45.

129. See <http://www.pbs.org/weta/thewest>.

130. Daniel Max, "Time Warner's 'West' Flunks Synergy Test," *Variety,* 2 June 1992, 1, 84; Ward, *The West: An Illustrated History.*

131. " *The West:* A Documentary Film Series," NEH grant application, 22.

132. " *The West:* A Documentary Film Series," NEH grant application, cover sheet.

133. Flanagan, "The Other Burns Comes Back East: With *The Way West.*"

CHAPTER 6

1. Ken Burns, response to a question during the Historians Film Committee panel "Television and History: Ken Burns's *Thomas Jefferson,*" at the 112th Annual Meeting

of the American Historical Association, 10 January 1998, Seattle, Washington.

2. National Commission on Excellence in Education, *A Nation at Risk: The Imperative for Educational Reform* (Washington, D.C.: U.S. Government Printing Office, 1983).

3. Lynne V. Cheney, *American Memory: A Report on the Humanities in the Nation's Public Schools* (Washington, D.C.: U.S. Government Printing Office, 1987); Michael Frisch, "American History and the Structures of Collective Memory: A Modest Exercise in Empirical Iconography," *Journal of American History* 75.4 (1989), 1131.

4. See Diane Ravitch, "Decline and Fall of Teaching History," *New York Times Magazine*, 17 November 1985; Simon Schama, "Clio Has a Problem," *New York Times Magazine*, 8 September 1991; Paul Gray, "Whose America?" *Time*, 8 July 1991; and Paul Gagnon, "Why Study History?" *Atlantic Monthly*, November 1988.

5. Caroline Waxler, "Move Over, Murrow," *Forbes*, 15 November 1999, 64; Richard Katz, "Bio Format Spreads Across Cable Webs," *Variety*, 2-8 August 1999, 23, 27; Kim McAvoy, "Prime Time for Documentaries," *Broadcasting & Cable*, 26 June 1998, 55-56; Richard Mahler, "Reality Sites," *Hollywood Reporter*, nonfiction special issue, 8 April 1997, N8-N9; and Ginia Bellafante, "These Are Their Lives," *Time*, 17 March 1997, 67.

6. JoAnna Baldwin Mallory, "Introduction," in *Telling the Story: The Media, the Public, and American History*, ed. Sean Dolan (Boston: New England Foundation for the Humanities, 1994), vii, ix.

7. Gerald Herman, "Chemical and Electronic Media in the Public History Movement," *Public Historian: A Journal of Public History* 21.3 (summer 1999), 113.

8. Herman, "Chemical and Electronic Media," 125.

9. Joseph J. Ellis, *American Sphinx: The Character of Thomas Jefferson* (New York: Knopf, 1997), 22.

10. Robert Sklar, "Historical Films: Scofflaws and the Historian-Cop," *Reviews in American History* 25.3 (1997), 346-50.

11. See Gary Edgerton, "Ken Burns's Rebirth of a Nation: Television, Narrative, and Popular History," *Film & History* 22.4 (1992), 118-33.

12. See Robert B. Toplin, ed., *Ken Burns's The Civil War: Historians Respond* (New York: Oxford, 1996). The seven responses to the series actually cover a spectrum of opinion ranging from complimentary (i.e., C. Vann Woodward and Robert B. Toplin) to ambivalent (i.e., Gabor S. Boritt and Gary W. Gallagher) to condemning (i.e., Eric Foner and Leon F. Litwack) and even dismissive (i.e., Catherine Clinton). The collection ends with the series' chief writer Geoffrey Ward and Ken Burns offering their own sides of the story, as well as their impressions about the chasm that all too often exists between themselves, the general public, and professional historians.

13. Catherine Clinton, "Noble Women as Well," in *Ken Burns's The Civil War: Historians Respond*, 66, 189.

14. Christopher Shea, "Taking Aim at the 'Ken Burns' View of the Civil War," *Chronicle of Higher Education*, 20 March 1998, A16.

15. James M. McPherson, *Battle Cry of Freedom* (New York: Oxford, 1988).

16. "Primetime TV Rate Race," *Hollywood Reporter* 366.4 (February 26, 1997), 37; Larry Bonko, "Ken Burns Series Reveals Virginia Roots of Famous Trek," *Virginian-Pilot*, 4 November 1997, E2.

17. Ken Burns quoted in "Process + Production" on the *Thomas Jefferson* (1997) website at <http://www.pbs.org/jefferson/making/KB_01.htm>.

18. "Special with Ken Burns on the Making of *Thomas Jefferson*," *Virginia Currents*, WHRO TV-15, Hampton Roads Public Television, 2 April 1997.

19. Ken Burns, speech, "Searching for Thomas Jefferson," at the Williamsburg Lodge, Colonial Williamsburg, Virginia, 4 February 1997.

20. Edgerton, "Ken Burns's Rebirth of a Nation," 130.

21. "*Thomas Jefferson*, a New Film by Ken Burns About One of Our Nation's Most Eloquent Presidents, Will Be Broadcast on PBS Feb. 18 and 19, 1997," *Thomas Jefferson Press Kit* (New York: Owen Comora Associates, a Division of Serino Coyne Public Relations, 1996), 1.

22. "*Thomas Jefferson*, a New Film," *Thomas Jefferson Press Kit*, 1.

23. "Ken Burns and the Historical Narrative on Television," Museum of Television & Radio University Satellite Seminar Series (90 minutes), 19 November 1996.

24. Ellis, *American Sphinx*, 3.

25. See Dumas Malone, *Thomas Jefferson and His Times*, 6 vols. (Boston: Little, Brown, 1948-1981); and Merrill D. Peterson, *The Jefferson Image in the American Mind* (New York: Oxford, 1960).

26. Burns, speech, "Searching for Thomas Jefferson."

27. Burns, speech, "Searching for Thomas Jefferson."

28. "Special with Ken Burns on the Making of *Thomas Jefferson*," *Virginia Currents*.

29. "Ken Burns Creates '18th Century Virtual Reality' in his Cinematic Portrait of Thomas Jefferson," *Thomas Jefferson Press Kit*, 2.

30. Eric Rudolph, "*Thomas Jefferson* Evokes Era of Enlightenment," *American Cinematographer* 78:1 (January 1997), 84.

31. "Ken Burns Creates '18th Century Virtual Reality,'" *Thomas Jefferson Press Kit*, 2.

32. B. Ruby Rich, "Documentarians: State of Documentary," *National Forum: The Phi Kappa Phi Journal* 77.4 (fall 1997), 23. The producer-directors taking part in the discussion were St. Clair Bourne, Arthur Dong, Rob Epstein, Su Friedrich, Deborah Hoffmann, Steve James, Ross McElwee, Errol Morris, Michel Negroponte, Lourdes Portillo, Renee Tajima-Pena, Jessica Yu, and Terry Zwigoff.

33. Ken Burns, interview, 27 February 1996.

34. Ken Burns, "Process + Production," *Thomas Jefferson* (1997) website.

35. The transcript of Burns's entire interview with Franklin can be found on the *Thomas Jefferson* (1997) website.

36. Ken Burns, speech, "Sharing the American Experience," Norfolk Forum, Norfolk, Virginia, 27 February 1996. See previous reference in note 23 to "Ken Burns and the Historical Narrative," Museum of Television & Radio, 19 November 1996.

37. Burns, interview, 27 February 1996.

38. Conversation between Ken Burns and the author prior to the Historians Film Committee panel, "Television and History: Ken Burns's *Thomas Jefferson*," at the 112th Annual Meeting of the American Historical Association, 10 January 1998, Seattle, Washington.

39. "Ken Burns: Exploring the Complexities and Inner Turmoil of Thomas Jefferson," *Thomas Jefferson Press Kit*, 1.

40. The timeline under "The Process" on the *Thomas Jefferson* (1997) website delineates that preparation began in spring 1990 and assembly ended during the fall of 1996.

41. Jack McLaughlin, *Jefferson and Monticello: The Biography of a Builder* (New York: Henry Holt, 1988), 383.

42. *Thomas Jefferson* was a production of Florentine Films in association with WETA-TV, Washington, D.C. Corporate funding was provided by General Motors Corporation; additional funding by the Pew Charitable Trusts, the Corporation for Public Broadcasting, the Public Broadcasting Service, Virginia Department of Tourism, and the Arthur Vining Davis Foundations.

43. Joseph J. Ellis, "Whose Thomas Jefferson Is He Anyway?" *New York Times,* 16 February 1997, H35.

44. Ken Burns as quoted in "Ken Burns: Uncovering American History," interview by Sharon McRill, Borders.com video editor, at <http://go.borders.com/fcgi-bin/part?PID'12...Eborders%Ecom%2Ffeatures%2Fskm99033%2Excv>.

45. Ken Burns, in "Filmmakers: Meet Ken Burns and Lynn Novick," interview by Eric Yeater, May 1998, on the *Frank Lloyd Wright* (1998) website at <http://www.pbs.org/flw/filmmakers/filmmakers_qa.html>.

46. From a 1990 curriculum vitae of Burns under the section, "Films by Ken Burns," in the resources on *The Civil War,* Ken Burns Collection at the Folklore Archives of the Wilson Library, University of North Carolina at Chapel Hill.

47. Burns, interview, 27 February 1996.

48. Ken Burns, "A Forum with Ken Burns," on the *Lewis & Clark: Journey of the Corps of Discovery* (1997) website at <http://www.pbs.org/lewis and clark/forum/intro.html>. See also Dayton Duncan, *Out West: American Journey Along the Lewis and Clark Trail* (New York: Penguin, 1988).

49. Ken Burns as quoted in McGrill, "Ken Burns: Uncovering American History." See Elisabeth Griffith, *In Her Own Right: The Life of Elizabeth Cady Stanton* (New York: Oxford, 1985).

50. See Geoffrey C. Ward, *Before the Trumpet: Young Franklin Roosevelt, 1882-1905* (New York: Perennial, 1986); and Geoffrey C. Ward, *A First-Class Temperament: The Emergence of Franklin Roosevelt* (New York: Harper & Row, 1989).

51. From a 1990 curriculum vitae of Burns under the section, "Films by Ken Burns," in the resources on *The Civil War,* Ken Burns Collection.

52. Jim Zwick, "The Ken Burns Documentary on Mark Twain: An Interview with Coproducer Dayton Duncan" December 1998, on the Mark Twain.About.Com website at <http://marktwain.miningco.com/arts/marktwain/library/weekly/aa990126.htm>.

53. Burns, interview, 27 February 1996.

54. Ken Burns quoted in "General Motors Mark of Excellence Presentations" on the General Motors website at <http://www.gm.com/company/sponsored_by_gm/pbs/index.htm>.

55. Budgetary figures in the *Empire of the Air* materials, Ken Burns Collection.

56. Conversation between Ken Burns and the author prior to the Historians Film Committee panel, "Television and History: Ken Burns's *Thomas Jefferson.*"

57. Ken Burns quoted in "General Motors Mark of Excellence Presentations" on the General Motors website.

58. Ken Burns quoted in Sara Switzer, "Third-Degree Burns," *George,* November 1999, 65.

59. Ken Burns, telephone interview with the author, 18 February 1993.

60. James Poniewozik, "Biosphere," *Time,* 23 August 1999, 62-66.

61. Katz, "Bio Format Spreads Across Cable Webs," 27; Poniewozik, "Biosphere," 63-64; Mahler, "Reality Sites," N9; and "'Biography' at 10," special advertising section, *Broadcasting & Cable,* 10 March 1997, S1-S20.

62. Bellafante, "These Are Their Lives," 67.

63. Bellafante, "These Are Their Lives," 67.

64. Poniewozik, "Biosphere," 63-64, 66; and "'Biography' at 10," S1-S20.

65. Debbie Messina, "Jefferson's Virginia: State Banks on Burns' Film Drawing Tourists," *Virginian-Pilot* (Norfolk), 5 February 1997, D1.

66. David Nicholson, "Williamsburg Plays Big Role in Ken Burns' 'Jefferson,'" *Daily Press* (Newport News), 5 February 1997, 15.

67. The nine historic sites included in "Jefferson's Virginia" are Colonial Williamsburg, College of William and Mary, Tuckahoe Plantation, State Capitol, Barboursville Ruins and Vineyards, Monticello, University of Virginia, Jefferson's Poplar Forest, and Natural Bridge.

68. Zwick, "Ken Burns Documentary on Mark Twain"; Messina, "Jefferson's Virginia: State Banks," D2.

69. Robert B. Toplin, "Plugged In to the Past," *New York Times,* sec. 2, 4 August 1996, 26.

70. "Tale of Twain," *Newsweek.com,* 30 December, 1998 at <http://www.newsweek.com/nw-srv/issue/01_99a/tnw/today/nm/nm01tu2_1.htm>.

71. "Whose Mark Twain?" *Mark Twain Newsletter,* 18 January 1999, 1, at <http://marktwain.about.com/arts/marktwain/library/newsletters/bl_news990118.htm>.

72. Zwick, "The Ken Burns Documentary on Mark Twain."

73. Ken Burns speech, "Searching for Thomas Jefferson"; see Robert Penn Warren, *The Legacy of the Civil War,* reprint edition (Lincoln: University of Nebraska Press, 1998).

74. Ken Burns, speech, "Searching for Thomas Jefferson."

75. See Meryle Secrest, *Frank Lloyd Wright: A Biography,* reprint edition (Chicago: University of Chicago Press, 1998).

76. Carolyn Anderson, "Biographical Film," *Handbook of American Film Genres,* ed. Wes D. Gehring (Westport, Conn.: Greenwood, 1988), 331.

77. Anderson, "Biographical Film," 332.

78. Ken Burns, interview by David Thelen, "The Movie Maker as Historian: Conversations with Ken Burns," *Journal of American History* 81.3 (1994), 1033.

79. "Burns Makes History at PBS" in "General Motors Mark of Excellence Presentations" on the General Motors website.

80. "'Lewis and Clark'—Ken Burns, Dayton Duncan," *Charlie Rose Show* (#2022), airdate: 3 November 1997.

81. Ken Burns quoted in "Filmmakers: Meet Ken Burns and Lynn Novick" on the *Frank Lloyd Wright* (1998) website at <http://www.pbs.org/flw/filmmakers/filmmakers.html>.

82. "Documentarians Describe Film on Frank Lloyd Wright—Ken Burns, Lynn Novick," *Charlie Rose Show* (#2287), airdate: 9 November 1998.

83. "Documentarians Describe Film on Frank Lloyd Wright," *Charlie Rose Show.*

84. Paul Barnes, in "The Making of 'Not For Ourselves Alone,'" PBS promotional special, airdate: 7 November 1999.

85. Burns, interview, 27 February 1996.

86. B. J. Bullert, *Public Television: Politics & the Battle over Documentary Film* (New Brunswick, N.J.: Rutgers University Press, 1997), 180.

87. Burns, "Filmmakers: Meet Ken Burns and Lynn Novick," *Frank Lloyd Wright* (1998) website.

88. Novick, "Filmmakers: Meet Ken Burns and Lynn Novick," *Frank Lloyd Wright* (1998) website.

89. Burns, "Filmmakers: Meet Ken Burns and Lynn Novick," *Frank Lloyd Wright* (1998) website.

90. Burns, "Filmmaker Ken Burns," PBS promotional special, airdate: 11 November 1998.

91. Burns, "Filmmakers: Meet Ken Burns and Lynn Novick," *Frank Lloyd Wright* (1998) website.

92. Burns, "Filmmakers: Meet Ken Burns and Lynn Novick," *Frank Lloyd Wright* (1998) website.

CHAPTER 7

1. Letter from Geoffrey C. Ward to Ken Burns, 9 January 1996, in "*Jazz:* A Documentary Film Series for Public Television Series," grant application submitted to the National Endowment for the Humanities, 10 January 1996, 70-71.

2. F. Scott Fitzgerald, *The Last Tycoon: An Unfinished Novel* (New York: Charles Scribner's Sons, 1970), 163.

3. Gary Giddins, "Jazz and America: An Interview of Geoffrey C. Ward," *American Heritage,* December/January 2001, 64.

4. Giddins, "Jazz and America," 64.

5. "*Jazz:* A Documentary Film Series," NEH grant application, 19.

6. Letter from Ward to Burns, 9 January 1996, in "*Jazz:* A Documentary Film Series," NEH grant application, 70.

7. Don Heckman, "Not Exactly All That Jazz," *Los Angeles Times,* Sunday Calendar, 6 August 2000, 9.

8. Ken Burns quoted in "Filmmakers: Meet Ken Burns and Lynn Novick" on the *Frank Lloyd Wright* (1998) website at <http://www.pbs.org/flw/filmmakers/filmmakers.html>.

9. "*Jazz:* A Documentary Film Series," NEH grant application, 20.

10. Peter Watrous, "Telling America's Story Through America's Music," *New York Times,* 1 October 2000, Arts & Leisure, sec. 2, 39-40.

11. "*Jazz:* A Documentary Film Series," NEH grant application, cover sheet and budget, 20.

12. "*Jazz:* A Documentary Film Series," NEH grant application, 22.

13. Madison D. Lacy had been associated with public television since 1972, first as a programming executive, then as an executive producer of cultural affairs programming at WGBH in Boston, and most recently as an independent producer. He won an Emmy for cowriting and producing *The Time Has Come* (1990) as part of the *Eyes on the Prize II* series. Peter Miller also had worked in various producing capacities over the previous decade. He coproduced *The Uprising of '34* (1995) and *Passin' It On* (1993), which were both shown nationally on PBS's *P.O.V.* series. He was coordinating producer for Barbara Kopple's Academy Award–winning documentary *American Dream* (1990); and he also worked for Florentine Films coordinating music clearances for *Baseball* and as an associate producer on the home video and Sony recording *Songs of the Civil War*

(1991). In addition to Marsalis, the advisory board was comprised of Michael Chertok, James Lincoln Collier, Stanley Crouch, Michael Cuscuna, Dayton Duncan, Julie Dunfey, Gerald Early, Tom Evered, Gary Giddins, Matt Glaser, Joanna Groning, Eric Hobsbaum, Robin D. G. Kelley, Charlie Lourie, Allen Lowe, Albert Murray, Daniel Okrent, Bruce Boyd Raeburn, Loren Schoenberg, Gunther Schuller, and Margaret Washington.

14. Letter from Wynton Marsalis to Ken Burns, 9 June 1995, in *"Jazz:* A Documentary Film," NEH grant application, 85.

15. Heckman, "Not Exactly All That Jazz," 9.

16. *"Jazz:* A Documentary Film Series," NEH grant application, 41.

17. Heckman, "Not Exactly All That Jazz," 9.

18. Ken Burns quoted in "Behind the Scenes: Interview with Ken Burns" on the *Jazz* (2000) website at <http://www.pbs.org/jazz/about/about_behind_the_scenes2.htm>.

19. Ken Burns, telephone interview with the author, 18 February 1993.

20. Claude Lévi-Strauss, *The Savage Mind* (Chicago: University of Chicago Press, 1966), 17. See also Claude Lévi-Strauss, *Totemism* (Boston: Beacon Press, 1963) and *The Raw and the Cooked* (New York: Harper & Row, 1969).

21. Lévi-Strauss, *The Savage Mind,* 21.

22. Lynn Novick quoted in "Behind the Scenes: Interview with Lynn Novick" on the *Jazz* (2000) website.

23. Ken Burns, "Searching for Thomas Jefferson," speech delivered at the Williamsburg Lodge, Colonial Williamsburg, Virginia, 4 February 1997.

24. Novick, "Behind the Scenes: Interview with Lynn Novick," *Jazz* (2000) website; "Acclaimed Film Director Ken Burns Celebrates the 'Most American of Art Forms' from Its Origins in Blues and Ragtime through Swing, Bebop and Fusion," *Jazz Press Kit* (New York: Dan Klores Associates, 2000), 2.

25. *"Jazz:* A Documentary Film Series," NEH grant application, 10.

26. Burns, "Behind the Scenes: Interview with Ken Burns." *Jazz* (2000) website; Watrous, "Telling America's Story," 40.

27. Jon Pult, "Backtalk (Interview) with Ken Burns," *OffBeat* (New Orleans' and Louisiana's music magazine), May 2000, at <http://www.offbeat.com/ob2005/back-talk.html>.

28. Tennessee Williams, *The Glass Menagerie* (New York: New Directions Books, 1970), 21.

29. Pult, "Backtalk (Interview) with Ken Burns."

30. Ken Burns is a particularly apt example of a filmmaker as bricoleur, although this label is not meant to also imply the full structuralist assumption that he (as all humans according to this outlook) orders the full range of his experiences into a series of binary oppositions. Interestingly, though, Burns and Ward often evoke polar opposites as a rhetorical strategy in their offscreen narrations to poetically capture the pluralism of American culture and society.

31. Letter from Madison D. Lacy to Ken Burns, 9 January 1996, in *"Jazz:* A Documentary Film Series," NEH grant application, 75.

32. A few jazz series were attempted, such as *Music '55* (CBS, July-September 1955) starring Stan Kenton and his orchestra, although all were short lived. Jazz has been used successfully as theme and background music on a number of television programs, however; some of the more memorable examples include *Peter Gunn* (NBC, 1958-

1960 and ABC 1960-1961), *Route 66* (CBS, 1960-1964), and *77 Sunset Strip* (ABC, 1958-1964).

33. Emily DeNitto, "Trumpeting Jazz: Ken Burns Focuses on Marketing His Film," *Crain's: New York Business,* October 16-22, 2000, 88.

34. Two examples of these objections are available on the web by Tom Cunniffe, "Ken Burns' 'Jazz,'" at <http://www.jazzinstituteofchicago.org/jazzgram/reviews/cunniffe-burns.asp>; and manager, agent, and producer Marty Khan, "An Open Letter to Ken Burns," at <http://www.birdlives.com/burns.html>.

35. Giddins, "Jazz and America," 65.

36. Geoffrey C. Ward, *Jazz: A History of America's Music,* with a preface by Ken Burns (New York: Knopf, 2000), 432.

37. Novick, "Behind the Scenes: Interview with Lynn Novick," *Jazz* (2000) website.

38. Burns, "Behind the Scenes: Interview with Ken Burns," *Jazz* (2000) website.

39. Burns, "Behind the Scenes: Interview with Ken Burns," *Jazz* (2000) website.

40. Watrous, "Telling America's Story," 40.

41. Ken Burns, interview by David Thelen, "The Movie Maker as Historian: Conversations with Ken Burns," *Journal of American History* 81.3 (1994), 1043.

42. David Gates, "The Story of Jazz," *Time,* 8 January 2001, 61.

43. Novick, "Behind the Scenes: Interview with Lynn Novick," *Jazz* (2000) website.

44. Burns, "Behind the Scenes: Interview with Ken Burns," *Jazz* (2000) website.

45. Pult, "Backtalk (Interview) with Ken Burns."

46. "*Jazz:* A Documentary Film Series," NEH grant application, 12.

47. Giddins, "Jazz and America," 65.

48. Burns, Behind the Scenes: Interview with Ken Burns," *Jazz* (2000) website.

49. Laurence Bergreen, *Louis Armstrong: An Extravagant Life* (New York: Broadway Books, 1997), 333.

50. Andrew Dansby, "From Armstrong to 'Zip-a-dee-do-da': Ken Burns' Jazz 101," *RollingStone On-Line,* 22 December 2000, at <http://dailynews.yahoo.com.../from_armstrong_to_zip-a-dee-do-da_ken_burns_jazz_101_1.htm>.

51. This section of Marsalis's filmed interview was incorporated by Ward into *Jazz: A History of America's Music,* 459.

52. Giddins, "Jazz and America," 67.

53. Ken Burns, interview with the author, 27 February 1996.

54. Ken Burns, quoted in "The Indie Scene: Producer's Interview," September 1998, at <http://www.pbs.org/independents/forum/sept98_forum.2.html>.

55. Ken Burns, "Sharing the American Experience," speech delivered at the Norfolk Forum, Norfolk, Virginia, 27 February 1996, 16.

56. Ken Burns quoted in "Filmmakers: Interview with Ken Burns and Paul Barnes" on the *Not For Ourselves Alone: The Story of Elizabeth Cady Stanton and Susan B. Anthony* (1999) website at <http://www.pbs.org/stantonanthony/filmmakers/filmmakers.html>.

57. Burns, "Filmmakers: Interview with Ken Burns and Paul Barnes," *Not For Ourselves Alone: The Story of Elizabeth Cady Stanton and Susan B. Anthony* (1999) website.

58. Mark Muro, "Mr. Civil War: Shelby Foote Makes His Presence Felt in PBS Epic," *Boston Globe,* 26 September 1990, 50.

59. Krin Gabbard, "Ken Burns's 'Jazz': Beautiful Music, but Missing a Beat," *Chronicle of Higher Education,* 15 December 2000, B18-B19.

60. Burns, "Behind the Scenes: Interview with Ken Burns," *Jazz* (2000) website.
61. Bill Kirtz, "Filmmaker's Riff on all that 'Jazz,'" *Christian Science Monitor*, 8 December 2000, 15.
62. Lévi-Strauss, *The Savage Mind*, 20-21.
63. Joanne Ostrow, "Ken Burns Searches for America's Soul in 'Jazz,'" *Denver Post*, 2 May 2000, 22.
64. *1998 Report on Television* (New York: Nielsen Media Research, 1998), 16.
65. Data available on the front page of the TV-Free America website at <www.rtis.com/reg/bcs/pol/touchstone/Summer1996/worsham.htm>.
66. Sheldon Hackney, "A Conversation with Ken Burns on Baseball," *Humanities* 15.4 (1994), 48-49.
67. Joel L. Swerdlow, "The Power of Writing," *National Geographic* 196.2 (1999), 124.
68. Burns speech, "Sharing the American Experience."
69. "David McCullough," *Current Biography* 54.1 (1993), 386.
70. Pult, "Backtalk (Interview) with Ken Burns."
71. Burns in Thelen, "The Movie Maker as Historian," 1049.
72. Lévi-Strauss, *The Savage Mind*, 21.
73. DeNitto, "Trumpeting Jazz," 88.
74. Lawrie Mifflin, "TV Notes; A 10-Year Bet on Ken Burns," *New York Times*, 16 June 1999, E10.
75. Robert Sullivan, "Visions of Glory," *Life*, September 1994, 44.
76. William Jefferson Clinton, "Remarks at a Screening of Ken Burns' 'Lewis and Clark,'" *Weekly Compilation of Presidential Documents* 33.46 (1997), 1782.
77. Burns, interview, 27 February 1996.
78. Dansby, "From Armstrong to 'Zip-a-dee-do-da.'"
79. "Ken Burns' 'Jazz' Doubles PBS Audience," (Reuters), *Yahoo! News*, 11 January 2001 at <http://dailynews.yahoo.com/h/nm/20010111/re/tv_jazz_dc_1.html>; "Nielsen Ratings," *Variety*, January 15-21, 2001, 44; "Burns' 'Jazz' Gives Music a Boost," (AP), *Yahoo! News*, 1 February 2001 at <http://dailynews.yahoo.com/h/ap/20010201/en/jazz_surge_1.html>; and Chris Morris, "Advent of a New Jazz Age?" *Billboard*, 10 February 2001, 1, 82.
80. Ben Ratliff, "Fixing, For Now, The Image Of Jazz," *New York Times*, 7 January 2001, Arts and Leisure, sec. 2, 32; "Burns' 'Jazz' Gives Music a Boost."
81. The 22 single-artist collections that Columbia/Legacy and the Verve Music Group combined their catalogs to produce included overviews of Louis Armstrong, Count Basie, Sidney Bechet, Art Blakey, Dave Brubeck, Ornette Colemen, John Coltrane, Miles Davis, Duke Ellington, Ella Fitzgerald, Dizzy Gillespie, Benny Goodman, Herbie Hancock, Coleman Hawkins, Fletcher Henderson, Billie Holiday, Charles Mingus, Thelonius Monk, Jelly Roll Morton, Charlie Parker, Sonny Rollins, and Lester Young.
82. Paul Robicheau, "Ken Burns Brand Sparks Leap in 'Jazz' Sales," *Billboard.Com*, 22 January 2001, at <http://daily news.yahoo.com/h/so/20010122/e....rns_brand_sparks_leap_in_jazz_sales_1.html>; "Top Jazz Albums," *Billboard*, 3 March 2001, 32.
83. Paul Robicheau, "Ken Burns Brand Sparks Leap."
84. Chris Morris, "Verve, Columbia/Legacy Team on 'Jazz' Series," *Billboard*, 9 September 2000, 18.

85. DeNitto, "Trumpeting Jazz," 87.

86. "General Motors Announces Integrated Marketing Plan to Support PBS Documentary on Jazz by Award-Winning Ken Burns," *Jazz Press Kit,* 3-4.

87. "General Motors Announces Integrated Marketing Plan," *Jazz Press Kit,* 2.

88. Barry A. Jeckell, "Marketing 'Jazz': A Collaborative Effort," *DailyMusicNews,* 6 September 2000, at <http://www.billboard.com/daily/2000/0906_08.as>

89. DeNitto, "Trumpeting Jazz," 88.

90. Burns, "The Indie Scene: Producer's Interview."

91. "Ken Burns and the Historical Narrative on Television," Museum of Television & Radio University Satellite Seminar Series (90 minutes), 19 November 1996.

92. Ratliff, "Fixing, For Now, The Image Of Jazz," 32.

93. Ward, *Jazz: A History of America's Music,* 459; and Tim Page, "In Avant-Guarded Condition: What's Ahead for Jazz? Mostly, What's Behind," *Washington Post,* 7 January 2001, Arts, G1.

94. Ward, *Jazz: A History of America's Music,* 460.

95. Ward, *Jazz: A History of America's Music,* 460.

96. Khan, "An Open Letter to Ken Burns."

97. Jay Tolson, "Stormy Weather: Debates over Jazz Rage on, as PBS Readies its New Series," *U.S. News & World Report,* 11 December 2000, 62; Watrous, "Telling America's Story," 40.

98. Burns, "Behind the Scenes: Interview with Ken Burns," *Jazz* (2000) website.

99. Morris, "Verve, Columbia/Legacy Team," 18.

100. Gabbard, "Beautiful Music, but Missing a Beat," B19.

101. Burns, interview, 27 February 1996.

102. Burns, "The Indie Scene: Producer's Interview."

103. Mifflin, "A 10-Year Bet on Ken Burns," E10.

Videography

Brooklyn Bridge (PBS, 1982); A Film by Ken Burns; Associate Producers: Amy Stechler, Buddy Squires, and Roger Sherman; Narrated by David McCullough; Cinematography by Ken Burns; Written and Edited by Amy Stechler; Additional Cinematography by Buddy Squires; Production Manager: Roger Sherman; Production of Florentine Films in Association with Thirteen/WNET, New York; 58 minutes; B & W and Color.

The Shakers: Hands to Work, Hearts to God (PBS, 1985); A Film by Ken Burns and Amy Stechler Burns; Narrated by David McCullough; Cinematography by Ken Burns and Terry Hopkins; Written by Amy Stechler Burns, Wendy Tilgham, and Tom Lewis; Production Associates: Wendy O'Connell, Toby Shimin, Kate Lynch, Peter Agoos, Peter Tagiuri, Roger Sherman, and Camilla Rockwell; Production Consultant: Buddy Squires; Production of Florentine Films in Association with the Television Laboratory at Thirteen/WNET, New York; 58 minutes; B & W and Color.

The Statue of Liberty (PBS, 1985); A Film by Ken Burns; Produced by Buddy Squires and Ken Burns; Narrated by David McCullough; Edited by Paul Barnes; Cinematography by Buddy Squires, Terry Hopkins, and Ken Burns; Written by Bernard A. Weisberger and Geoffrey C. Ward; Additional Writing by Ric Burns and Amy Stechler Burns; Production Manager: Roger Sherman; Production Associate: Camilla Rockwell; Assistant Editor: Bruce Shaw; Production of Florentine Films in Association with Thirteen/WNET, New York; 58 minutes; B & W and Color.

Huey Long (PBS, 1986); A Film by Ken Burns; Produced by Ken Burns and Richard Kilberg; Written by Geoffrey C. Ward; Edited by Amy Stechler Burns; Cinematography by Ken Burns and Buddy Squires; Narrated by David McCullough; Associate Producer/Writing: Ric Burns; Assistant Editor: Wendy O'Connell; Assistant Producer: Camilla Rockwell; Post-Production Supervisor: Buddy Squires; Sound by Greg Moring and Roger Sherman; RKB/Florentine Films Production; 88 minutes; B & W and Color.

Thomas Hart Benton (PBS, 1989); A Film by Ken Burns; Produced by Ken Burns and Julie Dunfey; Written by Geoffrey C. Ward; Edited by Donna Marino; Narrated by Jason Robards; Cinematography by Ken Burns and Buddy Squires; Script Consultant: Henry Adams; Assistant Editors: Tim Ballantine and Jean Marie Offenbacher; Associate Producer: Camilla Rockwell; Production of Florentine Films in Association with the WGBH Educational Foundation, Boston; 86 minutes; B & W and Color.

The Congress: The History and Promise of Representative Government (PBS, 1989); A Film by Ken Burns; Produced by Ken Burns and Stephen Ives; Written by David McCullough and

Bernard Weisberger; Edited by Sally Jo Menke; Narrated by David McCullough; Cinematography by Ken Burns, Alan Moore, and Buddy Squires; Associate Producers: Mike Hill and Catherine Eisele; Production Coordinator: Camilla Rockwell; Additional Writing: Geoffrey C. Ward and Ric Burns; Production of Florentine Films in Association with WETA, Washington, D.C.; 88 minutes; B & W and Color.

Lindbergh: The Shocking, Turbulent Life of America's Lone Eagle (PBS, 1990); Produced and Directed by Stephen Ives; Written by Geoffrey C. Ward; Narrated by Stacy Keach; Co-Producer: Ken Burns; Edited by Juliet Weber; Associate Producers: Michael Kantor and Bruce Alfred; Director of Photography: Buddy Squires; Music by Mark Bennett; Assistant Editor Rachel Warden; Additional Cinematography: Ken Burns, Mead Hunt, and Allen Moore; Insignia Films Production for *The American Experience* in Association with the WGBH Educational Foundation, Boston, and Thirteen/WNET, New York; 58 minutes; B & W and Color.

The Civil War (PBS, 1990); A Film by Ken Burns; Produced by Ken Burns and Ric Burns; Written by Geoffrey C. Ward and Ric Burns with Ken Burns; Narrated by David McCullough; Edited by Paul Barnes, Bruce Shaw, and Tricia Reidy; Cinematography by Ken Burns, Allen Moore, and Buddy Squires; Coordinating Producer: Catherine Eisele; Associate Producer/Post-Production: Lynn Novick; Coproducers: Stephen Ives, Julie Dunfey, and Mike Hill; Senior Creative Consultant: David McCullough; Associate Producers: Camilla Rockwell and Susanna Steisel; Production of Florentine Films in Association with WETA, Washington, D.C.; GM Mark of Excellence Presentation; 670 minutes; B & W and Color.

Empire of the Air: The Men Who Made Radio (PBS, 1992); A Film by Ken Burns; Produced by Ken Burns, Morgan Wesson, and Tom Lewis; Written by Geoffrey C. Ward; Edited by Paul Barnes; Associate Editor: Yaffa Lerea; Narrated by Jason Robards; Cinematography by Ken Burns, Buddy Squires, and Allen Moore; Associate Producers: Camilla Rockwell and Susanna Steisel; Additional Writing: Tom Lewis and Ken Burns; Based on the book, *Empire of the Air* (HarperPerennial, 1991) by Tom Lewis; Production of Florentine Films in Association with WETA, Washington, D.C.; GM Mark of Excellence Presentation; 120 minutes; B & W and Color.

Baseball (PBS, 1994); A Film by Ken Burns; Produced by Ken Burns and Lynn Novick; Written by Geoffrey C. Ward and Ken Burns; Supervising Film Editor: Paul Barnes; Edited by Paul Barnes, Yaffa Lerea, Tricia Reidy, Michael Levine, and Rikk Desgres; Coordinating Producers: Bruce Alfred and Mike Hill; Narrated by John Chancellor; Cinematography: Buddy Squires, Ken Burns, and Allen Moore; Assistant Editors: Erik Ewers and Matt Landon; Associate Producers: David Schaye and Susanna Steisel; Consulting Producer: Stephen Ives; Postproduction Associate: Kevin Kertscher; Production Manager: Camilla Rockwell; Production of Florentine Films in Association with WETA, Washington, D.C.; GM Mark of Excellence Presentation; 1110 minutes; B & W and Color.

Vézelay: Exploring the Question of Search with William Segal (1996); A Film by Ken Burns with Buddy Squires, Roger Sherman, and Wendy Conquest; Edited by Sarah Hill; Production of American Documentaries, Inc.; 31 minutes, Color. *Vézelay* includes a companion short entitled, *William Segal* (1992), 13 minutes. Another 30-minute film by

Burns, *In the Marketplace* (2000), further examines the perspective and art of philosopher and painter William Segal.

The West (PBS, 1996); A Film by Stephen Ives; Written by Geoffrey C. Ward and Dayton Duncan; Produced by Stephen Ives, Jody Abramson, and Michael Kantor; Senior Producer: Ken Burns; Supervising Film Editor: Paul Barnes; Edited by Richard Hankin, Michael Levine, and Adam Zucker; Narrated by Peter Coyote; Music by Matthias Gohl; Director of Photography: Buddy Squires with Allen Moore; Consulting Producer: Dayton Duncan; Associate Producer: Victoria Gohl; Script and Film Research: Michelle Ferrari; Production Manager: Suzanne Seggerman; Assistant Editors: Laura Congleton, George O'Donnell, Keir Pearson, and Jay Pires; Coproduction of Insignia Films and WETA, Washington, D.C. in Association with Florentine Films and Time-Life Video & Television; GM Mark of Excellence Presentation; 750 minutes; B & W and Color.

Thomas Jefferson (PBS, 1997); A Film by Ken Burns; Produced by Ken Burns and Camilla Rockwell; Written by Geoffrey C. Ward; Edited by Paul Barnes and Kevin Kertscher; Assistant Editor: Shannon Robards; Associate Producer: Susanna Steisel; Cinematography: Allen Moore, Ken Burns, Buddy Squires, Peter Hutton; Narrated by Ossie Davis; Voice of Thomas Jefferson: Sam Waterston; Production of Florentine Films in Association with WETA, Washington, D.C.; GM Mark of Excellence Presentation; 180 minutes; B & W and Color.

Lewis & Clark: The Journey of the Corps of Discovery (PBS, 1997); A Film by Ken Burns; Written by Dayton Duncan; Produced by Dayton Duncan and Ken Burns; Edited by Paul Barnes and Erik Ewers; Cinematography: Buddy Squires, Ken Burns, Allen Moore; Narrated by Hal Holbrook; Associate Producer: Susanna Steisel; Coordinating Producer: Pam Tubridy Baucom; Assistant Editor: Aaron Vega; Senior Creative Consultant: Geoffrey C. Ward; Production of Florentine Films in Association with WETA, Washington, D.C.; GM Mark of Excellence Presentation; 240 minutes; B & W and Color.

Frank Lloyd Wright (PBS, 1998); A Film By Ken Burns and Lynn Novick; Written by Geoffrey C. Ward; Produced by Ken Burns, Lynn Novick, and Peter Miller; Edited by Tricia Reidy; Narrated by Edward Herrmann; Cinematography: Buddy Squires and Ken Burns; Associate Producer: Shola Lynch; Associate Editor: Sarah Hill; Senior Creative Consultant: Geoffrey C. Ward; Coordinating Producer: Pam Tubridy Baucom; Additional cinematography: Allen Moore and Peter Hutton; Production of Florentine Films in Association with WETA, Washington, D.C.; GM Mark of Excellence Presentation; 153 minutes; B & W and Color.

Not For Ourselves Alone: The Story of Elizabeth Cady Stanton and Susan B. Anthony (PBS, 1999); A Film by Ken Burns and Paul Barnes; Directed by Ken Burns; Produced by Paul Barnes and Ken Burns; Written by Geoffrey C. Ward; Edited by Sarah E. Hill; Cinematography by Buddy Squires, Allen Moore, and Ken Burns; Narrated by Sally Kellerman; Associate Producer: Susanna Steisel; Senior Creative Consultant: Geoffrey C. Ward; Coordinating Producer: Pam Tubridy Baucom; Additional Cinematography: Roger Haydock and Tom Marini; Production of Florentine Films in Association with WETA, Washington, D.C.; GM Mark of Excellence Presentation; 210 minutes; B & W and Color.

Jazz (PBS, 2001); A Film by Ken Burns; Written by Geoffrey C. Ward; Produced by Ken Burns and Lynn Novick; Coproducers: Peter Miller and Victoria Gohl; Supervising Film Editor: Paul Barnes; Episode Editors: Paul Barnes, Sandra Marie Christie, Lewis Erskine, Erik Ewers, Sarah E. Hill, Craig Mellish, Shannon Robards, Tricia Reidy, and Aaron Vega; Narrated by Keith David; Cinematography: Buddy Squires and Ken Burns; Associate Producers: Sarah Botstein, Natalie Bullock Brown, and Shola Lynch; Consulting Producer: M. Davis Lacy; Senior Creative Consultant: Wynton Marsalis; Production of Florentine Films in Association with WETA, Washington, D.C.; GM Mark of Excellence Presentation; 1125 minutes; B & W and Color.

Mark Twain (PBS, 2002); Directed by Ken Burns; Written by Dayton Duncan and Geoffrey C. Ward; Produced by Dayton Duncan and Ken Burns; Edited by Erik Ewers and Craig Mellish; Cinematography: Buddy Squires, Allen Moore, and Ken Burns; Narrated by Keith David; Voice of Mark Twain: Kevin Conway; Associate Producer: Susanna Steisel; Assistant Editor: Christine Rose Lyon; Post Production Associate: Margaret Shepardson-Legere; Senior Creative Consultant: Geoffrey C. Ward; Coordinating Producer: Pam Tubridy Baucom; Additional Cinematography: Roger Haydock; Production of Florentine Films in Association with WETA, Washington, D.C.; GM Mark of Excellence Presentation; 240 minutes; B & W and Color.

Selected Bibliography

PRIMARY AND SECONDARY SOURCES

ARCHIVAL SOURCE

The Ken Burns Collection at the Folklore Archives of the Wilson Library, University of North Carolina at Chapel Hill contains footage (including outtakes) from his first nine films through *Baseball,* all of his scripts (in multiple drafts), his interviews with various witnesses and scholars, notes on decision making, test narrations, some financial records, correspondence, other related data.

BURNS'S EARLY FILM WORK

Bondsville (1972), 22 minutes (director, cameraman).
A Certain Slant of Light (1977), 30 minutes (cameraman, production assistant).
Consider the Sea (1980), 30 minutes (cameraman).
Here Today . . . (1977), 120 minutes (director of photography).
In the Irish Tradition (1978), 12 minutes (cameraman).
In Manhattan (1975), 60 minutes (soundman).
Invisible World (1980), 60 minutes (cameraman for one segment of this *Nova/Odyssey* production).
The Old Quabbin Valley (1981), 30 minutes (cameraman, editing consultant).
The Something of Harry Lott (1976), 10 minutes (producer, director).
Springfield Station from the Moon (1973), 7 minutes (cameraman).
Transfusion (1974), 10 minutes (producer, editor).
Truth About the Turkey (1978), 60 minutes (cameraman for one segment of this BBC production).
Two Families (1973), 22 minutes (cameraman, editor).
Working in Rural New England (1976), 27 minutes (producer, director, cameraman, editor).
Yan's 400 Foot Movie (1974), 12 minutes (cameraman).

GRANT APPLICATIONS

"*Baseball:* Documentary Film Series for Public Television," 133-page grant application submitted to the National Endowment for the Humanities, 7 March 1991.
"Brooklyn Bridge Film Project," grant application submitted to the National Endowment for the Humanities, 28 August 1978, 48 pages and supporting appendices.

"*The Civil War:* A Television Series," 72-page grant application with appendices submitted to the National Endowment for the Humanities, 14 March 1986.

"Every Man a King: Huey Long and the Rise of Media Politics," 42-page grant application submitted to the National Endowment for the Humanities, 26 May 1982.

"*Jazz:* A Documentary Film Series for Public Television Series," 118-page grant application with cover sheet and budget submitted to the National Endowment for the Humanities, 10 January 1996.

"Radio Pioneers," 64-page grant application submitted to the National Endowment for the Humanities, 17 March 1989.

"*Thomas Hart Benton:* An American Original," 137-page grant application submitted to the National Endowment for the Humanities, 12 March 1986.

"*The West:* A Documentary Film Series for Public Television," 160-page grant application submitted to the National Endowment for the Humanities, 2 September 1992.

PUBLISHED INTERVIEWS

Burns, Ken. Interviewed in "The Indie Scene: Producer's Interview." September 1998, at <http://www.pbs.org/independents/forum/sept98_forum.2.html>.

Cripps, Thomas. "Historical Truth: An Interview with Ken Burns." *American Historical Review* 100.3 (June 1995): 741-64.

Dansby, Andrew. "From Armstrong to 'Zip-a-dee-do-da': Ken Burns' Jazz 101." *RollingStone On-Line,* 22 December 2000, at <http://dailynews.yahoo.com.../ from_armstrong_to_zip-a-dee-do-da_ken_burns_jazz_101_1.htm>.

"Documentarians Describe Film on Frank Lloyd Wright—Ken Burns, Lynn Novick," *Charlie Rose Show* (#2287), airdate: 9 November 1998.

Edgerton, Gary. "Ken Burns—A Conversation with Public Television's Resident Historian." *Journal of American Culture* 18.1 (1995): 1-12.

Gerzon, Mark and Molly De Shong. "E Pluribus Unum: Ken Burns and the American Dialectic." *Shambhala Sun,* November 1997 at <http://www.shambhalasun.com/ Archives/Features/1997/Nov97/KenBurns.htm>.

Giddins, Gary. "Jazz and America: An Interview of Geoffrey C. Ward." *American Heritage,* December/January 2001, 63-70.

Hackney, Sheldon. "A Conversation With Ken Burns on *Baseball.*" *Humanities* 15, July/ August 1994: 4-7, 48-53.

"Historian Goodwin on Baseball Memoir," *Charlie Rose Show* (#2063), airdate: 30 December 1997.

"Ken Burns and 'Baseball' Talk." *Charlie Rose Show* (#1211), airdate: 23 September 1994.

Lalire, George. "Interview: *The West* According to Burns and Ives." *Wild West,* October 1996, at <http://www.thehistorynet.com/WildWest/articles/1096_text.htm>.

"'Lewis and Clark'—Ken Burns, Dayton Duncan." *Charlie Rose Show* (#2022), airdate: 3 November 1997.

Pult, Jon. "Backtalk (Interview) with Ken Burns." *OffBeat* (New Orleans' and Louisiana's Music Magazine), May 2000, at <http://www.offbeat.com/ob2005/back-talk.html>.

Spanberg, Erik. "Commentary: Ken Burns on *Civil War,* Clinton, Success." *The Business Journal of Charlotte,* 1 March 1999, at http://www.amcity.com:80/charlotte/stories/1999/03/01/editorial3.html>.

Thelen, David. "The Movie Maker as Historian: Conversations with Ken Burns." *Journal of American History* 81.3 (December 1994): 1031-50.

"'The Way West'—Ric Burns." *Charlie Rose Show* (#1371), airdate: 5 May 1995.

"'The West'—Ken Burns Documentary." *Charlie Rose Show* (#1723), airdate: 10 September 1996.

Wiener, David Jon. "Interview with 1998 Golden Laurel Recipient: Producer Ken Burns." *Point of View On-Line* 1.1 (1999) at <http://www.empire-pov.com/burns.html>.

Zwick, Jim. "The Ken Burns Documentary on Mark Twain: An Interview with Coproducer Dayton Duncan." December 1998, on the Mark Twain.About.Com website at <http://marktwain.miningco.com/arts/marktwain/library/weekly/aa990126.htm>.

PRESS KITS

Baseball Press Kit (New York: Owen Comora Associates, Div. of Serino Coyne Public Relations, 1994).

The Civil War Press Kit (New York: Owen Comora Associates, Div. of Serino Coyne Public Relations, 1990).

Jazz Press Kit (New York: Dan Klores Associates, 2000).

Lewis & Clark Press Kit (New York: Owen Comora Associates, Div. of Serino Coyne Public Relations, 1997).

Thomas Jefferson Press Kit (New York: Owen Comora Associates, Div. of Serino Coyne Public Relations, 1996).

The West Press Kit (New York: Owen Comora Associates, Div. of Serino Coyne Public Relations, 1996.)

SPEECHES

Burns, Ken. "The Documentary Film: It Role in the Study of History." Text of speech delivered as a Lowell Lecture at Harvard College, 2 May 1991, 12 pages.

———. "Jerome Liebling as Teacher." Text of speech delivered at Hampshire College, 31 March 1990, 5 pages.

———. "Mystic Chords of Memory." Text of speech delivered at University of Vermont, 12 September 1991, 19 pages.

———. Remarks at the "Lewis Mumford at 100" Symposium. Meyerson Hall, University of Pennsylvania, 5 October 1995 (author's audiotaped copy).

———. "Searching for Thomas Jefferson." Text of speech delivered at Williamsburg Lodge, Colonial Williamsburg, Virginia, 4 February 1997, 21 pages.

———. "Sharing the American Experience." Text of speech delivered in Norfolk, Virginia, 27 February 1996, 30 pages.

National Press Club. "Ken Burns, Documentary Filmmaker." Transcript of 12 September 1994 appearance. Washington, D.C.: C-Span, 1994, 15 pages.

VIDEOTAPES

"A Conversation with Filmmaker Ken Burns," Part II. WHRO TV-15, Hampton Roads Public Television, PBS Promotional Special (10 minutes), airdate: 5 November 1997.

"Ken Burns and the Historical Narrative on Television." Museum of Television & Radio University Satellite Seminar Series (90 minutes), 19 November 1996.

"Ken Burns's Jefferson." The 112th American Historical Association Conference, Seattle, Washington, 10 January 1998, C-SPAN 1 (107 minutes; ID#98568), rebroadcast 25 May 1998.

"The Making of *Baseball*." WHRO TV-15, Hampton Roads Public Television (30 minutes), 4 July 1994.

"The Making of 'Not For Ourselves Alone.'" WHRO TV-15, Hampton Roads Public Television, PBS Promotional Special (10 minutes), airdate: 7 November 1999.

"Special with Ken Burns on the Making of *Thomas Jefferson*." *Virginia Currents*, WHRO TV-15, Hampton Roads Public Television (30 minutes), 2 April 1997.

Working in Rural New England (1976). Produced and directed by Ken Burns in association with Old Sturbridge Village, 27 minutes.

WEBSITES

Not For Ourselves Alone: The Story of Elizabeth Cady Stanton and Susan B. Anthony (1999) website at <http://www.pbs.org/stantonanthony>.

Frank Lloyd Wright (1998) website at <http://www.pbs.org/flw>.

Jazz (2001) website at <http://www.pbs.org/jazz>.

Lewis & Clark (1997) website at <http://www.pbs.org/lewisandclark>.

Thomas Jefferson (1997) website at <http://www.pbs.org/jefferson>.

The West (1996) website at <http://www.pbs.org/thewest>.

GENERIC NEWSPAPERS AND MAGAZINES
(CONTAINING REVIEWS AND NEWS UPDATES)

Billboard
Broadcasting & Cable
Daily Variety
Electronic Media
Hollywood Reporter
The New York Times
Nielsen Newscast
TV Guide
Variety
Washington Post

BOOKS

Barnouw, Erik. *Documentary: A History of the Non-Fiction Film.* 2nd rev. ed. New York: Oxford University Press, 1993.

Browne, Nick, ed. *American Television: New Directions in History and Theory.* Langhorne, Pa.: Harwood Academic Publishers, 1994.

Bullert, B. J. *Public Television: Politics and the Battle over Documentary Film.* New Brunswick, N.J.: Rutgers University Press, 1997.

Burgoyne, Robert. *Film Nation: Hollywood Looks at U.S. History.* Minneapolis: University of Minnesota Press, 1997.

Burns, Amy Stechler and Ken Burns. *The Shakers—Hands to Work, Hearts to God: The History and Visions of the United Society of Believers in Christ's Second Appearing from 1774 to the Present.* Photographs by Ken Burns, Langdon Clay, Jerome Liebling, and from Shaker archives, and foreword by Elderess Bertha Lindsay. New York: Aperture, 1987.

Burns, Ken. "Foreword," in *Walking to Cold Mountain: A Journey Through Civil War America.* Carl Zebrowski. New York: Smithmark, 1999: 8-10.

Cantor, Muriel. *The Hollywood TV Producer: His Work and His Audience.* New York: Basic Books, 1971.

Carnes, Mark C., ed. *Past Imperfect: History According to the Movies.* New York: Henry Holt, 1995.

Cullen, Jim. *The Civil War in Popular Culture: A Reusable Past.* Washington, DC: Smithsonian Institution Press, 1995.

Day, James. *The Vanishing Vision: The Inside Story of Public Television.* Berkeley: University of California Press, 1995.

Dolan, Sean, ed. *Telling the Story: The Media, The Public, and American History.* Boston: New England Foundation for the Humanities, 1994.

Edgerton, Gary, Michael T. Marsden, and Jack Nachbar, eds. *In the Eye of the Beholder: Critical Perspectives in Popular Film and Television.* Bowling Green State University Press, 1997 (in Chapter 1—"'Mystic Chords of Memory': The Cultural Voice of Ken Burns": 11-26).

Edgerton, Gary R., and Peter C. Rollins, eds. *Television Histories: Shaping Collective Memory in the Media Age.* Lexington: University Press of Kentucky, 2001.

Ellis, Joseph J. *American Sphinx: The Character of Thomas Jefferson.* New York: Knopf, 1997.

Engelman. Ralph, *Public Radio and Television in America: A Political History.* Thousand Oaks, Cal.: Sage, 1996.

Ferncase, Richard K. *Outsider Features: American Independent Films of the 1980s.* Westport, Conn.: Greenwood, 1996.

Foote, Shelby. *Civil War: A Narrative (Fort Sumter to Perryville, Fredericksburg to Meridan, Red River to Appomattox),* 3 vols. New York: Random House, 1958-1974.

Fussell, Paul. *The Great War and Modern Memory.* New York: Oxford, 1989.

Gans, Herbert J. *Popular Culture & High Culture: An Analysis and Evaluation of Taste.* Rev. ed. New York: Basic Books, 1999.

Heider, Karl G., ed. *Images of the South: Constructing a Regional Culture on Film and Video.* Athens: University of Georgia Press, 1993 (in Chapter 7—"Was It Not Real? Democratizing Myth Through Ken Burns's *The Civil War*": 112-123).

Jarvik, Laurence. *PBS: Behind the Screen.* New York: Prima, 1998.

Kammen, Michael. *Mystic Chords of Memory: The Transformation of Tradition in American Culture.* New York: Vintage, 1993.

Landy, Marcia. *Cinematic Uses of the Past.* Minneapolis: University of Minnesota Press, 1996.

———, ed. *The Historical Film: History and Memory in Media.* New Brunswick, N.J.: Rutgers University Press, 2001.

Ledbetter, James. *Made Possible By: The Death of Public Broadcasting in the United States.* New York: Verso, 1998.

Le Goff, Jacques. *History and Memory,* European Perspectives. New York: Columbia University Press, 1996.

Levine, Lawrence W. *The Unpredictable Past: Explorations in American Cultural History.* New York: Oxford, 1993.

Lewis, Bernard. *History: Remembered, Recovered, Invented.* Princeton, N.J.: Princeton University Press, 1975.

Lewis, Tom, *Empire of the Air: The Men Who Made Radio.* New York: HarperPerennial, 1993.

Liebling, Jerome. *The People, Yes.* "Foreword" by Carroll T. Hartwell. New York: Aperture, 1995.

Liebling, Jerome. *Jerome Liebling Photographs.* With essays by Anne Halley and Alan Trachtenberg. Amherst: University of Massachusetts Press, 1982.

Lipsitz, George. *Time Passages: Collective Memory and American Popular Culture.* Minneapolis: University of Minnesota Press, 1990.

Marc, David, and Robert J. Thompson. *Prime Time, Prime Movers: From I Love Lucy to L.A. Law—America's Greatest TV Shows and the People Who Created Them.* Boston: Little, Brown, 1992 (in Chapter 29—"Ken Burns: The Art of the Artifact": 301-308).

Marsden, Michael T., John G. Nachbar, and Sam L. Grogg, Jr. *Movies as Artifacts: Cultural Criticism of Popular Film.* Chicago: Nelson-Hall, 1982.

McArthur, Colin. *Television and History.* London: British Film Institute, 1978.

McLaughlin, Jack. *Jefferson and Monticello: The Biography of a Builder.* New York: Henry Holt, 1988.

McCullough, David, *The Great Bridge: The Epic Story of the Building of the Brooklyn Bridge.* New York: Touchstone, 1972.

Newcomb, Horace, and Robert S. Alley. *The Producer's Medium.* New York: Oxford University Press, 1983.

Novick, Peter. *That Noble Dream: The "Objectivity Question" and the American Historical Profession.* Cambridge: Cambridge University Press, 1988.

O'Connor, John E., ed. *American History/American Television: Interpreting the Video Past.* New York: Ungar, 1983.

———. *Image as Artifact: The Historical Analysis of Film and Television.* Malabar, Fla.: Krieger, 1990.

———. *Teaching History With Film and Television,* Rev. ed. New York: American Historical Association, 1988.

O'Connor, John E., and Martin A. Jackson, eds. *American History/American Film: Interpreting the Hollywood Image.* New York: Ungar, 1979.

Rollins, Peter C., ed. *Hollywood as Historian: American Film in Cultural Context.* Lexington: University Press of Kentucky, 1983.

Rose, Brian G. *Directing for Television: Conversation with American TV Directors.* Metuchen, N.J.: Scarecrow, 1999.

Rosenstone, Robert A. *Revisioning History: Film and the Construction of a New Past.* Princeton, N.J.: Princeton University Press, 1995.

Schudson, Michael. *Watergate in American Memory: How We Remember, Forget and Reconstruct the Past.* New York: Basic Books, 1992.

Shaara, Michael, *The Killer Angels.* New York: Ballantine, 1974.

Smith, Paul, ed. *The Historian and Film.* Cambridge: Cambridge University Press, 1976.

Sobchack, Vivian, ed. *The Persistence of History: Cinema, Television and the Modern Event.* New York: Routledge, 1996.

Sontag, Susan. *On Photography.* New York: Doubleday, 1978.

Susman, Warren. *Culture as History: The Transformation of American Society in the Twentieth Century.* New York: Pantheon, 1984.

Thompson, Robert J. *Adventures on Prime-Time: The Television Programs of Stephen J. Cannell.* New York: Praeger, 1990.

————, and Gary Burns, eds. *Making Television: Authorship and the Production Process.* New York: Praeger, 1990.

Toplin, Robert B. *History by Hollywood: The Use and Abuse of the American Past.* Urbana: University of Illinois Press, 1996.

————, ed. *Ken Burns's The Civil War: Historians Respond.* New York: Oxford, 1996.

————, ed. *Perspectives on Audiovisuals in the Teaching of History.* Washington D.C.: American Historical Association, 1999.

Vidal, Gore. *Screening History.* Reissue edition. Cambridge, Mass.: Harvard University Press, 1994.

Ward, Geoffrey C., and Ken Burns, *Baseball: An Illustrated History.* New York: Knopf, 1994.

————, with Ric Burns and Ken Burns. *The Civil War: An Illustrated History.* New York: Knopf, 1990.

————, *Jazz: A History of America's Music.* With a preface by Ken Burns. New York: Knopf, 2000.

————, Ken Burns, and Jim O'Connor. *Shadow Ball: The History of the Negro Leagues (Baseball, the American Epic).* New York: Random House, 1994.

————, Ken Burns, and S. A. Kramer. *25 Great Moments (Baseball, the American Epic).* New York: Random House, 1994.

————, *The West: An Illustrated History.* With a preface by Stephen Ives and Ken Burns. Boston: Little, Brown, 1996.

————, Ken Burns, and Paul Robert Walker. *Who Invented the Game (Baseball, the American Epic).* New York: Random House, 1994.

Warren, Robert Penn. *All the King's Men.* New York: Harvest, 1996. (Originally published in 1946.)

————, *The Legacy of the Civil War,* Reprint edition. Lincoln: University of Nebraska Press, 1998.

White, Hayden. *Tropics of Discourse: Essays in Cultural Criticism.* Baltimore: Johns Hopkins University Press, 1978.

Zelizer, Barbie. *Covering the Body: The Kennedy Assassination, the Media, and the Shaping of Collective Memory.* Chicago: University of Chicago Press, 1992.

Zielinski, Siegfried. *Audiovisions: Cinema and Television as Entr'actes in History.* Translated by Gloria Custance. Amsterdam: Amsterdam University Press, 1999.

ARTICLES

Blight, David W. "Homer with a Camera, Our 'Iliad' without the Aftermath: Ken Burns's Dialogue with Historians." *Reviews in American History* 25:2 (1997): 351-59.

Bode, Ken. "Hero or Demagogue? The Two Faces Of Huey Long on Film." *New Republic,* 3 March 1986: 37-41.

Bösel, Anke. "Ken Burns's Film Series 'The Civil War:' An Attempt at American Self-Definition." *American Studies* [Germany] 40:2 (1995): 283-89.

Burns, Ken. "In Search of the Painful, Essential Images of War." *New York Times,* 27 January 1991, sec. 2: 1.

Burns, Robert K., Jr. "Saint Véran," *National Geographic* 115.4 (1959): 572-88.

Censer, Jane Turner. "Videobites: Ken Burns's 'The Civil War' in the Classroom." *American Quarterly* 44.2 (1992): 244-54.

"*The Civil War:* Ken Burns Charts a Nation's Birth." *American Film,* 15 (September 1990): 58.

Cohn, William H. "History for the Masses: Television Portrays the Past." *Journal of Popular Culture* 10 (1976): 280-89.

Cornfield, Michael. "What is Historic about Television?" *Journal of Communication* 44 (1994): 106-16.

"David McCullough." *Current Biography* 54.1 (1993): 384-87.

DeCredico, Mary A., "Image and Reality: Ken Burns and the Urban Confederacy," *Journal of Urban History* 23.4 (1997): 387-405.

Duncan, Dayton. "A Cinematic Storyteller." *Boston Globe Magazine,* 19 March 1989: 19, 72, 77-83.

Early, Gerald. "American Integration, Black Heroism, and the Meaning of Jackie Robinson." *Chronicle of Higher Education,* 23 May 1997: B4.

Edgerton, Gary. "Ken Burns." *The Encyclopedia of Television News.* Ed. Michael D. Murray. New York: Oryx Press, 1999: 27-28.

———. "Ken Burns" (258-60) and "The Civil War" (373-75). *Encyclopedia of Television.* Ed. Horace Newcomb. Chicago: Fitzroy Dearborn, 1997.

———. "Ken Burns's America: Style, Authorship, and Cultural Memory." *Journal of Popular Film and Television* 21.2 (1993): 50-62.

———. "Ken Burns's American Dream—Histories-for-TV from Walpole, New Hampshire." *Television Quarterly* XXVII.I (1994): 56-64.

———. "Ken Burns's Rebirth of a Nation: Television, Narrative, and Popular History." *Film & History* 22.4 (1992): 118-33.

Freehling, William W. "History and Television." *Southern Studies* 22 (1983): 76-81.

Henderson, Brian. "*The Civil War:* 'Did It Not Seem Real?'" *Film Quarterly* 44.3 (1991): 2-14.

Herman, Gerald. "Chemical and Electronic Media in the Public History Movement." *Public Historian: A Journal of Public History* 21.3 (summer 1999): 111-25.

Jackson, Donald D., "Ken Burns Puts His Special Spin on the Old Ball Game," *Smithsonian* 25.4 (July 1994): 38-50.

"Ken Burns." *Biography Today* 4.1 (January 1995): 68-75.

"Ken Burns." *Current Biography,* May 1992: 6-10.

Ketch, Jack. "Touching All the Bases." *The World and I,* January 1995, 151-61.

Koeniger, A. Cash. "Ken Burns's 'The Civil War': Triumph or Travesty?" *Journal of Military History* 55 (April 1991): 225-33.

Lancioni, Judith. "The Rhetoric of the Frame: Revisioning Archival Photographs in *The Civil War." Western Journal of Communication* 60 (1996): 397-414.

Lester, Valerie. "Happy Birthday, Tom Benton!" *Humanities* 10.6 (1989): 32-33.

Lewis, Jan, and Peter S. Onuf. "American Synecdoche: Thomas Jefferson as Image, Icon, Character, and Self." *American Historical Review* 103:1 (1998): 125-36.

Lewis, Thomas S. W. "Radio Revolutionary: Edwin Howard Armstrong's Invention of FM Radio," *American Heritage of Invention & Technology,* fall 1985, 34-41.

Maslin, Janet. "Visionaries With Their Eyes on the Truth." *New York Times,* sec. 2 (2 May 1999): 37-38.

May, Robert E. "The Limitations of Classroom Media: Ken Burns' Civil War Series as a Test Case." *Journal of American Culture* 19.3 (1996): 39-49.

McFeely, William S. "Notes on Seeing History: the Civil War Made Visible." *Georgia Historical Quarterly* 74 (1990): 666-71.

McPherson, Tara. "'Both Kinds of Arms': Remembering *The Civil War." Velvet Light Trap* 35 (1995): 3-18.

Milius, John. "Reliving the War Between Brothers." *New York Times,* 16 September 1990, sec. 2: 1, 43.

O'Connor, John E. "The Moving Image Media in the History Classroom." *Film & History* 16 (1986): 49-54.

Rich, B. Ruby. "Documentarians: State of Documentary," *National Forum: The Phi Kappa Phi Journal* 77.4 (fall 1997): 20-25.

Rudolph, Eric. "*Thomas Jefferson* Evokes Era of Enlightenment." *American Cinematographer* 78:1 (January 1997): 81-84.

Sklar, Robert. "Historical Films: Scofflaws and the Historian-Cop," *Reviews in American History* 25 (June 1997): 346-50.

Slotkin, Richard. "'What Shall Men Remember?': Recent Work on the Civil War." *American Literary History* 3 (1991): 120-35.

Sullivan, Robert. "The Burns Method: History or Myth-Making?" *Life,* September 1994, 42.

———. "Visions of Glory." *Life,* September 1994, 40-5.

Thomson, David. "History Composed with Film." *Film Comment* 26.5 (September/October 1990): 12-16.

Tibbetts, John C. "The Incredible Stillness of Being: Motionless Pictures in the Films of Ken Burns." *American Studies* 37.1 (Spring 1996): 117-33.

Toplin, Robert B., "Plugged In to the Past," *New York Times,* sec. 2 (4 August 1996): 1, 26.

Zelizer, Barbie. "Reading the Past Against the Grain: The Shape of Memory Studies," *Critical Studies in Mass Communication* 12.2 (1995): 214-39.

General Index

Film/TV Index

Printed in the United States
by Bookmasters

Printed in the United States
By Bookmasters